Flash Forward

Flash Forward

A Series of Futuristic Vignettes

edited by

Nora Savage

Anita Street

PAN STANFORD PUBLISHING

Published by

Pan Stanford Publishing Pte. Ltd.
Penthouse Level, Suntec Tower 3
8 Temasek Boulevard
Singapore 038988

Email: editorial@panstanford.com
Web: www.panstanford.com

British Library Cataloguing-in-Publication Data
A catalogue record for this book is available from the British Library.

Flash Forward: A Series of Futuristic Vignettes

ISBN 978-981-4669-44-3 (Hardcover)
ISBN 978-981-4669-45-0 (eBook)

Printed in the USA

Contents

Introduction

Imagination is the inspiration of research, the fodder of children's daydreams, the spark of invention and creativity. So it should not be looked upon with surprise that this collection of science fiction stories is composed by science and engineering researchers. In addition, the field of nanotechnology has inspired innovative ideas and products; hence that the authors of this volume work in the science fiction arena should raise no eyebrows. However, what may be surprising to some is the clarity, the vision, and the amazingly interesting stories contained within this book.

Suspend for a moment your theorems and scientific proofs and enter a world where anything is possible, where anyone is capable, where time is merely another dimension or tool in the hands of the characters. Consequently, the plots, character thoughts, and actions take devious twists and curve round strange bends as the plots unfold.

Who says that the clever engineer cannot also be gifted artistically? Why should it be thought normal that the scientifically creative mind of the academician or researcher should never delve into the realm of science fiction for fear of being derided? Where is the objective proof that convinces us that the theoretician ought not to employ her or his creative juices to the development of a story as well as to the development of a hypothesis? Perhaps the still "siloed" halls of academia mandate this. Or possibly the constrained corporate structure. Or maybe the bureaucratic culture of government. Whatever the origin of this false barrier, its genesis is most certainly not found in the youth. The youth of which we speak is not one of years but of mentality or spirit. Nothing is impossible to the young of spirit or heart. Nothing is forbidden for the youthful of countenance.

Accordingly we bid you welcome, welcome to the stories of the future. We welcome you to the imaginative prospect of other dimensions. We welcome you if, as Sinatra stipulates, "If you are among the very young at heart."

Chapter 1

Ahead of Time

William Sims Bainbridge

On May 1, 1930, I fell from space toward northern Germany, on a sacred mission to kill the most dangerous man in the world. To conserve mass, my spacesuit and ceramic foam heat shield were built to the strictest specifications, and I had only 10 minutes of air. Petit woman that I am, the entry mass was well under a hundred kilograms. I felt the slightest pressure of the outermost atmosphere and braced myself for the frightful ride that was only just beginning. This was no time for second thoughts, but memory took me back to the moment, either 5 years ago or 500 years ahead, depending on one's perspective, when the desperate plan was hatched.

Most members of my generation are fanatics of one kind or another, but the style of fanaticism depends on the individual's character. I myself had gravitated toward one of the more obscure antidystopian movements, called the Clarkites. Some consider us a religious cult, but we are really a very rational organization dedicated to undoing the great mistake of the twentieth century.

Flash Forward: A Series of Futuristic Vignettes
Edited by Nora Savage and Anita Street

ISBN 978-981-4669-44-3 (Hardcover), 978-981-4669-45-0 (eBook)
www.panstanford.com

Practically everyone in the twenty-fifth century agrees that the world has descended into madness, but we disagree over the precise psychiatric diagnosis. Some say we have become immoral and corrupt, so they prescribe abstinence and prayer as the proper cures. Others say we are overstressed by the pressures of population and toxic wastes, so they prescribe sensuous pleasures and mood-lifting drugs. We Clarkites reject such simplistic solutions, finding our diagnosis and cure for the world's near-fatal ills in the writings of a twentieth-century genius, Arthur C. Clarke.

Clarke is best known for inventing the communication satellite, but he also was a prolific author of philosophy and prophetic fiction. His classic book *The Exploration of Space* is the nearest thing to a bible our movement possesses. There, he explained how spaceflight might begin, sometime in the twenty-first century, after the technology and economy of the Earth had evolved to the point that it could easily undertake voyages to the planets.

Unfortunately, and we Clarkites believe this was the crucial error that doomed the world, the Germans flew the first spaceship prematurely in 1942. The Americans reached the moon in 1969, far earlier than history had intended. Humanity was not ready for the universe, and so it recoiled from the great challenge, pretending for a few decades that it had a space program, while actually making no progress. Then, discredited by a few well-publicized fiascoes, human travel into space ended altogether early in the twenty-first century. Four hundred years later, I was the first person in a very long time to leave the atmosphere, and now only a few seconds later I was hurtling back down toward the Earth.

Imperfections in the heat shield had set me spinning but on an axis that kept the brilliant corona of flame away from my fragile suit. I could not get my bearings or even tell how fast I was spinning, because the flames filled my view and the deceleration overwhelmed my sense of balance. With great difficulty I touched the button that darkened my visor, and I read the clock and altimeter in my helmet. Only a few seconds and I would slow to the stratospheric terminal velocity. I touched the button again and watched the flames quiet to a reddening glow.

A moment later the tiny ribbon chute opened, jerking me free of the heat shield. I knew it was too early to open the main parachute, because I would freeze to death and run out of air if I stayed too high too long. So my thoughts drifted back to the scientific analysis and raw emotion that had brought me to this perilous moment, and I continued to fall.

In the issue of *Time* magazine commemorating the tenth anniversary of the first lunar landing, Clarke had put his point very succinctly: "Space travel is a technological mutation that should not really have arrived until the 21st century. But thanks to the ambition and genius of Wernher von Braun, the moon was reached half a century ahead of time." All we Clarkites had memorized those prophetic words. Because the first interplanetary voyage occurred ahead of time, the proper historical sequence was broken, and greater voyages were never undertaken. My mission was to get the evolution of space travel back on schedule.

Time travel is not impossible, only extremely difficult. To move a 40 kilogram woman from one year to another requires expending the total energy that would be liberated by completely converting her mass. A fission bomb converts only something like a tenth of a percent of the mass into energy, and a fusion bomb approaches one percent. But to drive me and my equipment back in time required the complete annihilation of one hundred kilograms of mass. This incidentally destroyed the entire island where our secret base was located, near the city of Palmer in West Antarctica. The world should be thankful we Clarkites had not built the launcher at our world headquarters in Sri Lanka. And I was personally thankful that the stasis field designed to hold my atoms together had worked, and so far my re-entry was proceeding safely.

I checked the instruments again, and saw that it was about time for the main chute. It opened automatically before I could trigger it by hand, with a rough jerk that slowed me to a few feet per second. I opened the helmet vents to let in some of the fresh German air. Below, the land was dark. To the east I could see the glow of impending sunrise.

Without the satellites of the Global Positioning System, I could not determine my exact location, but the geographic information display of my miniature digital library matched the scene below very closely, and I took the parachute shrouds in my hands to aim toward the clearing in the forest where I hoped to land. My mood was oddly split, as though I were simultaneously terrified and coolly confident. But, then, nobody had ever really understood my emotions, least of all myself.

My non-Clarkite friends must have thought I had left their century to escape neurotic depression from my last failed romance, after Harold joined the Castrati and subjected himself to their cruel initiation surgery. But, to me, my lover's foolish gesture was only one more sign of the world-poisoning pathology that must be cured. If I can fulfill my mission, then Harold would never have to leave me, or I would never have to join up with him in the first place. Success will change everything, including perhaps my fundamental personality, even my preferences in men. But that was only an incidental motive. My real reason for volunteering was the simple fact that I was the best person for a horribly difficult job.

As I tumbled safely on the soft grass of that clearing in the north German woods, exactly where I had intended to land, I felt renewed confidence I could complete my mission. I would kill the most dangerous man in the world, a young aristocrat named Wernher von Braun.

* * *

The sun had risen above the trees when Wernher rode his horse to point on the path where I was waiting. We knew from a letter he would write a cousin in a few days that he took a ride along this path every morning of this particular holiday. "Hallo!" I called to him. "Sir, would you possibly be so kind as to assist me?" I fingered the tiny capsule of sarin nerve gas that could end his life in a few seconds, but curiosity and caution prevented me from using it at that instant. I told him that I had been hiking, but twisted my ankle, so he graciously let me ride his horse for the hour it took us to reach the hotel where he and his father were staying.

Along the way we chatted, and this presented a considerable challenge to me, despite my extensive training in the dead language that he spoke. Although the old literature was still taught in some twenty-fifth century universities, German had become merely the "Deutsch" dialect of English, used only for intimate conversations and heavily loaded with non-German words and idioms. For example, in Wernher's era, German had a whole set of definite articles (*der, die, das, und so weiter*) that by my time had become simply *d′*, spelled like the English form, *the*.

I tried to explain to Wernher why I had been stranded in this lonely forest, conjuring up images of a deceitful elderly gentleman who had offered me a ride and then attempted to seduce me. But I could see from his expression that Wernher was perplexed by what I was saying. Then he asked the unexpected question, "Parlez-vous français?" When I pretended not to understand him, he continued, "Oh, pardon me. I was marveling at your very interesting accent and wondered if you might be French."

"No," I replied, "I am a German citizen, although my family lived for many years in the German districts of Czechoslovakia." This was the slightly hazardous fiction we had concocted to explain my strange style of speaking before I had a chance to hear ordinary German spoken and adapt to it. We had thought of locating my fictitious childhood in the German areas of Romania, but that was where Wernher's rocketry mentor, Hermann Oberth, had grown up, so we worried that he might expose my ignorance of the territory.

Most recorded examples of 1930s German that still existed in the twenty-fifth century were songs and motion pictures in which people spoke with very dramatic intonations. In 1929, even the science fiction film *Die Frau im Mond*, based on Oberth's ideas for a trip to the moon, had been a silent picture. My favorite was Marlene Dietrich's first sound movie, *The Blue Angel*, which premiered just a month before the day I met Wernher. I had (or from his temporal perspective *will have*) watched it over and over, and it may have given me some of the verbal habits of a cabaret vamp.

We arrived at the hotel just when Wernher was scheduled to meet his father for lunch, and I was rather ceremoniously introduced to Baron von Braun, who cordially invited me to join them. I could tell that he looked at me oddly, but I was not sure whether there was some subtle anachronism in my costume or something unfashionable about my style of movement. Possibly he categorized me as a gold digger, an unscrupulous woman who preyed on wealthy but innocent men. To be sure, he must have had difficulty identifying my proper place in the German social class hierarchy, but as a damsel in distress rescued by his son, I deserved provisional definition as a lady rather than a woman.

As we chatted over the luncheon, I kept looking for an opportunity to let Wernher discover that I shared his romantic vision of spaceflight, but he kept talking politely about things he imagined a young German lady of the day would be interested in. Perhaps it was a strain for him, because eventually he paused, then commented, "I think we live at a marvelous point in history, when so many new things are possible. For example, at the present state of science and technology, it is possible to build a machine that could rise higher than the Earth's atmosphere reaches."

He had jumped into the topic I wanted to discuss so suddenly that I replied rather too aggressively, "Yes, and you have just quoted the first sentence of a remarkable book by Hermann Oberth." Probably I should have coaxed him to tell me more, rather than admitting my deep knowledge of his favorite subject. Unfamiliar as I was with the subtle customs of early twentieth-century Germany, I was unsure how aggressive a woman should be if she wanted to attract a man.

Wernher seemed positively startled. "Yes! Certainly I did just utter the first words of *Die Rakete zu den Planetenräumen*. But how would you know such a thing?"

"Oh, but I have read the book, and I believe that space travel is possible. I gather you believe this as well. Nicht wahr, Herr von Braun?" I noticed a pained expression on the face of Wernher's father and sensed I was going too far.

The baron looked me coolly in the eye and spoke in a tone that combined conviction with a certain disdain. "My son has made himself an expert on these matters, and I am greatly pleased at the way he has taken control of his life, after a start that was not entirely promising. A father must respect the accomplishments of his son, even if they are far from his own field. In my life, I have seen many technical advancements, and I am convinced that Germany must lead the other nations technically as well as in the courage of its manhood. However, my son has the opportunity to achieve in many different fields, so what role he may play in the future of exploration beyond the atmosphere remains entirely undecided."

The conversation shifted to the sorry plight of the impoverished farmers in eastern Germany. After a dessert of *Pflaumenkompott*, which was too sweet for my tastes, the baron arranged for me to be transported to the railway station.

* * *

My training had prepared me for many contingencies, and we had a whole series of plans, some of which did not involve killing Wernher. Plan A expected me to act decisively on the first day, while other plans would attempt to halt spaceflight development weeks or even years later. I do not know why I did not use the sarin capsule at my first opportunity. Perhaps I simply did not have my bearings yet in the strange world of 1930. Maybe I failed to use it out of fear for my own life or because it was hard to feel that this eighteen-year-old boy really needed to die. In any case, I switched to plan B.

In Berlin, I checked into a modest but convenient hotel, where I emptied my money belt and began opening bank accounts and buying securities. Most of the currency I carried was large-denomination foreign notes, stolen by Clarkites from the museums of the twenty-fifth century, but there were a few low-denomination German Marks for immediate needs. I sent a letter to the *Verein für Raumschiffahrt* (VfR)—the Association for Spaceship Travel—along with a donation, applying to join. The VfR was the world's first spaceflight organization, the second having been founded at the birthday party of a science fiction editor in New York just weeks before my landing, and Wernher was among the VfR's nearly thousand members.

Since it was six months after the Wall Street crash, I was able to buy many securities at bargain prices, although the world economy had not yet reached the full depths of the Great Depression. The miniature digital library I carried had full information about which stocks would rise and fall over the next several decades, so I made my investments with perfect confidence. I was less confident about the stylish clothing I bought, because none of it felt right when I wore it. The digital library even gave me complete data about a failing businessman who needed to sell his year-old Mercedes-Benz SSK, which I wanted to carry out plan B.

Two weeks later, I drove the SSK onto the Opel test track, where Max Valier was demonstrating his latest rocket car. Although Oberth had received some publicity from *Die Frau im Mond*, Valier was by far the most famous rocket man in the world. He had created the first rocket-powered aircraft, which was a reasonable thing to do. But he also was responsible for the first rocket railway engine, which was simply silly. A wild romantic with a great flair for publicity but little technical sense, he had convinced the Opel car company to build a number of rocket cars. They were extremely hard to control and totally impractical for racing, let alone for any sane purpose.

Valier could not grasp Oberth's concept of multistage rocketry, and he imagined that spaceships would naturally evolve from aircraft as speeds and altitudes gradually increased, rather than requiring different design principles. However, Valier had just started experimenting with liquid-fuel engines, potentially more powerful and controllable than the solid-fuel rockets he had been working with.

If Valier continued in this direction, he would dominate German rocketry, making inferior technical decisions and thereby stalling spaceflight development, which was exactly what we Clarkites wanted him to do. Unfortunately, on May 17, 1930, Max Valier would be killed when an engine he was testing exploded. This left the field clear for Wernher and the other competent disciples of Oberth. So on the day before this disaster would occur, I dressed in my most spectacular sportswoman attire and drove my sports car up to where Valier was standing.

With Valier and his men watching my every move, I ostentatiously removed my driving goggles and cap, replacing them by a stylish feathered hat with the veil that elegant ladies of the day often wore to protect their pale faces from the sun. Valier himself opened the door of my SSK and expressed obvious pleasure when I asked him to show me his test auto. This he did in great detail, although the technical features of the car were quite obvious to anyone who looked, from the several Sander solid-fuel rockets in the rear to the crude switches on the control panel that would electrically fire each one.

Soon, Valier was fussing over me more than his car. I made him set up the two folding camp chairs, parasol, and tea table that had been occupying the SSK's trunk and second seat. Then we chatted about trivia while the test driver prepared for his brief but hazardous sprint. Valier asked me to give the signal, and a wave of my lace-trimmed handkerchief caused the first pair of rockets to blaze into action.

The car lurched forward, a great plume of smoke spewing from the rear. The driver had great difficulty with the first turn, his arms visibly straining to wrench the steering wheel around. The first pair of rockets quit, giving him a chance to brake while entering the second curve. Wisely he waited for the car to slow almost to halt at the beginning of the straightaway before igniting the second pair of rockets. Miraculously, he made it around the track twice more before coming to a full stop near the starting point, several pairs of rockets still unused. I waved my handkerchief before my face, to ward off some of the sulfurous fumes that were drifting our way, and effusively praised Valier's invention.

My plan had been to seduce Valier, delay him from testing his liquid-fuel engine, and find various ways to improve his workshop's safety measures. But he failed to react to my erotic hints, and I had to accept the fact that ladies of his era simply did not seduce men, unlike cabaret girls, who did so at every opportunity. I was barely able to coax from him an invitation to view his new engine the following day.

It was a crazy setup. Two bucket-like tanks would send propellants through a pair of pipes controlled by ordinary plumbing valves into the cylindrical engine. Valier poured a mixture of gasoline and water into one tank, the water hopefully serving to cool the engine, and steaming liquid oxygen into the other. He then lit a device rather like a firework sparkler near the exhaust and opened each valve by hand. Despite the water and the frigidity of the liquid oxygen, the sparkler ignited the mixture, and a great fan of fire issued from the engine. I knew, but Valier did not know, that the water would fail to cool the engine evenly. It would burn through at some point, unleashing holy hell.

The popular image of being knocked out by a blow to the skull is not at all like the reality. It is not a simple matter of switching consciousness off like a light and then switching it back on again. The last thing I remembered was Valier jostling me as he rushed around the rocket engine, marveling at the flame and checking the flow of propellants. I have no recollection of the explosion itself, but the nurses later told me that a fragment of the steel engine had struck my left temple. They must be right because I have scars there from the blow itself and the surgeon's work in removing a bone splinter from my brain. I have no memory of any pain.

The nurses also say that I raved about many foolish things, after I came out of the coma, but all I remember of those days was gradually becoming aware that a visitor often sat silently watching me. At some point I recognized it was Wernher, and gradually I became aware that he had learned of my accident and realized I was the only woman who was as passionate about the future of spaceflight as he was. Perhaps I loved him even then.

I was relieved to hear that Valier was unhurt, but my injury had so unnerved the man that he had abandoned work on liquid-fuel engines. The worsening economy made it impossible for Opel to invest in further publicity stunts with rocket cars, and Valier had become fascinated with the simple mechanical television devices that several inventors had built. So he soon became an evangelist for *picture radio*, touring the country to demonstrate a crude camera and projector set, each with a huge spinning disk punctuated with tiny holes that scanned the image, thick lenses to concentrate the light, and either a photoelectric tube or a neon bulb to take or make the picture. One of Wernher's associates, Arthur Rudolph, was dogging Valier's steps, trying to convince the visionary to turn over his plans and test data for the engine that had nearly killed me.

* * *

The hospital released me in late July. During my recuperation, I had begun thinking about crossing the Channel to England, to find my hero, Arthur C. Clarke, who was then a thirteen-year-old boy. But I realized that interfering with his life could ruin everything, so I resisted that powerful temptation. Instead, I immediately checked the safe deposit box where I had stashed my digital library, the three sarin capsules, and the other items that did not belong to the twentieth century, thankfully finding everything in good order. Plan C involved simply seducing Wernher, distracting him and thereby preventing him from accomplishing his first great political triumph in 1932. But I took one of the sarin capsules as a backup measure, bought a little silver box to keep it in, and henceforth carried it in my handbag.

I could tell that Wernher was fascinated by me, but as much as I flirted with him, he treated me more like an older cousin than a potential lover. He was an engineering student at the time, and we frequently met during the day for tea and occasionally went to a concert. One time we stayed in the concert hall after the performance, and he played Chopin beautifully on the Bechstein grand piano. Despite his strength of character and equanimity, I sensed he was still a bit shy among women. He never asked me my age, but on documents I always put 1903, which would make me nine years older than he.

It took me months to realize that plan C was failing. By this time, I was well established in German society, and could use some of my money to hire a man to do history's dirty work for me, following plan D.

Wernher was by nature a sportsman, and in the summer of 1932 he took gliding lessons at a flight training school in Silesia. This removed him from me not only geographically but emotionally, because there he acquired an aviator girlfriend named Hanna Reitsch, who would constantly fly in and out of his life for the next thirteen years.

I coached my agent carefully, and he was able to infiltrate the flight school, where he could arrange an accident to Wernher's glider. One of the standard methods for launching gliders was to tow them aloft behind a powered aircraft. There was a lever in the glider's cockpit that would release its end of the rope connecting the two craft, and then the powered aircraft could reel the rope in. My agent fixed Wernher's glider so that the lever would not work. It would be impossible for the two craft to land together. Thus, the other pilot would be forced to release his end of the rope, and it would destabilize the glider, leading to a crash that would certainly be fatal.

We did not reckon with Wernher's determination and ingenuity. The flight began normally as the powered aircraft hauled the glider aloft. But then came time to detach the rope. As soon as he realized the lever was broken, Wernher calmly searched the cockpit of his glider for things that could be used as tools. Familiar with the lever mechanism, he realized that pulling the rope back and to the left could release it, but only after the plane released its end to give the rope slack.

With the sharp edge of a piece of glass broken from his goggles, he cut a small opening in the doped fabric of the forward floor of the glider and then looped his belt around the rope. He signaled for the other pilot to release his end. Wernher's glider immediately went into a dive. He pulled hard on his belt and then let it and the rope fall away. In a few minutes he made a normal landing.

When my agent reported to me in Berlin, I was strangely elated to learn that Wernher had survived. The man wanted to try again, cutting Wernher's rudder cable so he would lose control, but I vetoed the idea. Over the next several years I employed him to spy on Wernher discretely, keeping me informed of his progress. There were several points during the 1930s when I could easily arrange to be alone with Wernher, far from any observers, and use one of the sarin capsules. Although I saw him every few months, I let those opportunities pass, telling myself that I would surely take action at the major opportunities that would come in 1943 and 1944.

* * *

History is a fabric of a trillion knotted threads, and cutting one seldom has any discernable effect. However, at certain points in the weave, many strands are bound together by a single thread. Cut it, and the whole fabric unravels. There were two such points in the history of the German space program. First, one of the rocket pioneers had to convince the military to invest heavily in rocket technology. Second, the technology had to be demonstrated spectacularly in the Second World War so that the Soviet Union and United States would be sufficiently impressed to begin their missile and space race.

My first several action plans, designated A through G, were all directed at preventing a rocket pioneer from taking the military on the correct detour from the battlefield to outer space. The failure of these plans allowed Wernher to take the first of these two decisive steps. He succeeded where much older men had failed, convincing the army to subsidize the rocket experiments of the bankrupt VfR. While still an engineering student in 1930, he had become one of the most active members of the VfR, testing crude liquid-fuel engines and giving public lectures. In 1932, he met Walter Dornberger, one of the new engineer-warriors of the German Army, and Colonel Karl Becker, who provided Wernher with a laboratory at the Kummersdorf proving ground. In December 1934, Wernher carried out successful demonstrations of small liquid-fuel rockets he called the A-2.

In the many Clarkite meetings leading up to my mission, there were great debates about what to do in any given year of the midtwentieth century in order to stall spaceflight development. A consensus had emerged that killing Wernher soon after the successful launches of the A-2 would be ineffective. Some argued that he had already assembled enough of a team to carry on without him or that the German Luftwaffe would see an opportunity to move ahead of the army, as both were by now investing in the work. My digital library reminded me that the Luftwaffe had recruited Dr. Eugen Sänger in 1936 to do research that led to designs for an intercontinental manned rocket aircraft, and removing Wernher from the picture might stimulate them to accelerate their support of this work several years earlier than they in fact did.

Furthermore, if Wernher died through obvious murder, the rage of the whole German military-industrial complex might pour resources into rocketry, under the assumption that British or French agents were responsible. The generals were suspicious of Wernher's teacher, Oberth, because while ethnically German, he was a citizen of Romania. But if they felt the British or the French were trying to sabotage their rocket program, they could very well place their confidence in Oberth. Thus, there was nearly a decade, from 1934 to 1942, when it seemed useless to kill Wernher. Therefore, plans H through L were canceled. But when this historical breathing space was used up, he might have to die.

So I invested the remainder of the 1930s preparing myself for the wave of opportunities that would come nearer the second great step in the history of spaceflight, when the A-4 rocket would be demonstrated in war. I carefully established an identity for myself and became the darling of the Schutzstaffel (SS) police.

German males of the period liked their women docile, but a handful of women became darlings of the Nazis precisely because they played against this stereotype. One was Leni Riefenstahl, who produced the stunning propaganda film *Triumph of the Will*, about Adolf Hitler's Nuremberg rallies. Another was Hanna Reitsch,

Wernher's girlfriend. She was the first person to fly a glider across the Alps, first to fly a helicopter inside a building, and later one of the test pilots for the Me-163 rocket interceptor. Wernher once told me, "Hanna is by far the most courageous and fearless girl I met in my life." He might not have said this had he known the truth about me.

Nobody, of course, realized that my political prominence came from the data in my digital library. My fellow SS officers merely thought I had an uncanny ability to ferret out information about disloyal citizens, hidden bank accounts, and plots against them personally, when really I was merely looking the information up in my digital library. I especially curried favor with the Nazi leaders who would play a role in Wernher's life: Albert Speer and Gerhard Degenkolb in the Ministry of Armaments, plus SS leaders Heinrich Himmler and Hans Kammler.

At the same time I did several apparently unrelated things that were actually preparing the way for the variety of contingency plans designated M through S. One of the most important was resuming violin practice. My skill with the violin was actually rare in the twenty-fifth century, because the decline of interest in classical music during the late twentieth and early twenty-first centuries had broken the historical chain of teachers and students who had maintained the quality of technique over the generations. After two years of intensive lessons unlearning my bad habits of bowing and fingering, I began playing with a serious amateur string quartet in Berlin.

* * *

In 1936, the army built a development facility for Wernher's rockets near the village of Peenemünde, on an island to the north of Berlin, where he could fire test missiles northward into the sea. At first I visited occasionally, merely because I was Wernher's friend, but then I took on a formal liaison assignment for the SS.

Technically, Wernher outranked me, because he held a commission as major in the SS. However, this was merely a politically necessary ceremonial position, and he seldom took his officer's uniform out of the closet. I, on the other hand, lacked a formal rank but was a civilian employee reporting directly to Himmler's staff office in Berlin. Later I was assigned to Kammler's staff when it began to take an interest in controlling developments at Peenemünde.

By the spring of 1942, I had my own office right down the hall from Wernher. He considered me a valuable asset, because I often cut through bureaucratic red tape for him, but of course he was unaware I was constantly looking for an opportunity to sabotage his work in a decisive manner.

I knew that the first two test flights of the massive A-4 rocket had been failures—or rather, would have been failures, assuming that nothing had changed the fate documented for them in the histories available to the Clarkite movement of the twenty-fifth century. It was always possible that some tiny change in events, brought about by my sheer presence, would let one of these test flights succeed. In the planning sessions for my time travel mission, we had debated whether I should sabotage the first two tests as insurance against such an eventuality. Those ideas were plan M and plan N. But we decided not to risk losing the chance to carry out plan O and sabotage the third test, which history said had been successful.

On June 13, 1942, the engine of the first test rocket fired, lifting the fifteenmeter-tall rocket off the launch stand, but then the thrust faltered. The A-4 settled back and fell over, rupturing the tanks containing the alcohol fuel and liquid oxygen, splashing fire all over the launch area.

A little over two months later, I stood beside Wernher as he watched the second A-4 climb slowly from its test stand and accelerate into the summer air. The expression on his face was a remarkable harmony of strained anticipation and spiritual transcendence, like a violin virtuoso bowing two strings on the instrument, each with a very different quality. I knew that he was especially worried about the control system, which depended upon graphite vanes in the exhaust, which might burn away before the engine reached *Brennschluss*, and about whether the fuselage could tolerate passing through the sound barrier.

Thirty seconds into the flight, it was clear the rocket had exceeded the speed of sound. But until it left the atmosphere and the engine cut off, we could not be sure it would survive the vibration and pressure. When the A-4 was about thirteen kilometers from launch, 45 seconds into the flight, it lost control and exploded.

I had expected Wernher to shout, or curse, or somehow reveal the disappointment he was feeling. But his face kept its intensely ambivalent expression. He closed his eyes and stood frozen for a long moment, like a still photograph of a man caught in an instant of deep reflection.

Then he spoke. "We cannot be sure which of several factors caused the failure. However, insufficient strength of the fuselage for the maximum air resistance is a possible factor. Whatever other changes we make, we must strengthen the midsection of the fuselage for the next test. Naturally, whenever we add strength, we also add weight. So if the next flight is successful, we will simplify the structure on later test vehicles judiciously in order to reduce weight and thereby gain range and payload."

Now the emotional pressure was on me. I needed to make sure that the third test would also fail. With my steadily increasing but secret wealth, I had indirectly subsidized a small communist cell in nearby Hamburg that had a master machinist among its members. Having a free run of Peenemünde, because of my SS connections and friendship with Wernher, I was able to steal one of the graphite vane assemblies and carry it away in the large leather shoulder bag I sometimes carried. It was not one of the four vanes designated for the upcoming test vehicle, so no one missed it from the supply of miscellaneous parts.

My communist machinist removed the metal bracket that would attach the vane to the servo mechanism that activated it, drilled several large holes in the back, wedged a small lead weight in each hole so the assembly would not seem too light, and then replaced the bracket. With the greatest difficulty I smuggled the part back into Peenemünde. Three times, I carried it in my shoulder bag as I walked past the parts rack where the four steering vanes belonging to the test rocket were carefully stored, until a moment came when no one was watching and I could switch the parts.

The morning of October 3, 1942, dawned clear and cool. I had planned to watch the launch from the athletic field where many of the common workers gathered, but a guard insisted that I follow him onto the roof of one of the camouflaged buildings, where Peenemünde's military commander, Walter Dornberger, was waiting with his pal, Leo Zanssen.

"Ah, glad you could come," said Dornberger. "From here we should have a fine view of the rocket once it rises above the trees, and we can see ignition on this little instrument." He indicated a small television device on which I could clearly make out the image of the A-4 on its launch stand. "Also, there is a matter of security that Colonel Zannsen wishes to discuss with you."

In recent months, I had many debates with Zannsen, because his main job for Dornberger and Wernher was fending off interference from Nazi officials, and my own ostensible function was representing the interests of the party's own army, the SS. Thus we often argued, and I felt sure he detested me and the organization I pretended to serve. What security matter could Zannsen have to discuss with me that would be on Dornberger's mind during the launch? Undoubtedly one having to do with the launch itself, quite possibly my sabotage of the rocket's steering vane!

"However," Dornberger continued, "I see that the countdown is about to begin." A green signal flare rose above the trees warning us that only 10 seconds remained. Dornberger handed me his binoculars, as his own attention riveted on the small television screen.

I raised the binoculars to my eyes, thankful for being able to hide my face while I struggled with the terror rising within me. For a moment I focused on the flare and then scanned the forest horizon until the red brick walls of the local cathedral came into view. The image shook wildly because my hands had begun to tremble. I gave the binoculars to Zannsen and noticed that his hands were quivering, too. Perhaps my fear would not make them suspicious, because they were as emotionally aroused as I was from the sheer excitement of the launch.

"Ignition!" shouted a voice over the loud speaker, and the two men echoed, "Ignition!" I knew the hydrogen-peroxide-powered pumps of the A-4 had begun to force alcohol and liquid oxygen into the combustion chamber, and I thought back to that day a dozen years earlier when I stood beside Valier as his crude engine exploded. The A-4's propellants would not ignite on contact, so I knew that an electric fuse had to touch off a small firework device. We saw the result on the tiny television screen, a blast of flame bathing the launch stand, overwhelming the television camera with light and obscuring the scene.

Dornberger reached for his binoculars, and Zannsen gave them over. With my naked eyes, I could see the pointed nose of the rocket rising majestically over the trees, as the roar of its engine bathed us in sound, like that of a huge waterfall. At this moment, I knew that the guidance of the missile depended entirely upon the graphite vanes, as the vehicle had not yet achieved enough speed for the four tailfins to grab the air. Would the vane I had sabotaged fail now, sending the vehicle into a catastrophic spin?

But the A-4 continued to rise straight and true. Dornberger stood ramrod straight himself, scrutinizing the flame spewing from the rocket for any hint of instability. As it rose higher and higher, he leaned back, in perfect synchronization with his machine. I could see it slowly turn and begin to angle northward, as its program intended, so that it would fly out over the sea.

As the roar lessened, I realized that Zannsen was counting seconds since the launch out loud. I could also hear a tone that was gradually changing in pitch, the Doppler transmitter from the A-4 amplified through the loudspeaker like electronic music expressing the rocket's rise. "Sonic velocity!" Zannsen called. We all knew that the second launch had reached this point and then failed shortly afterward. In my mind, I could imagine that graphite vane in the rocket's exhaust, burning through. In a moment, it might break away, and the control would become unbalanced.

But the tiny spot that was the rocket grew smaller and smaller into the blue sky and gradually disappeared. "Brennschluss!" The loudspeaker called. The engine had exhausted its fuel, far above the atmosphere, and now the A-4 was a pure missile following its parabolic trajectory in free fall. I took the binoculars and could just make out the tiny white speck that was the glow of the graphite vanes that had survived nearly a minute of intense fire.

"Thank heaven!" I exclaimed, oddly unsure whether I was really pleased my attempt at sabotage had failed or merely trying to express the emotion they would expect me to have.

"Yes," commented Zannsen, "but thanks are also deserved by our vigilant engineering staff. Not only have they accomplished a hitherto impossible technical feat, but they have also achieved a security success that really was not their job."

"What do you mean?" I inquired, trying not to appear apprehensive.

"Despite the guards your SS so kindly provided," Zannsen said with mock gratitude, "a sabotage attempt nearly succeeded. One involving a defective steering vane."

I allowed surprise to show on my face, expecting momentarily to be accused but hoping they would think I was angered to know that someone had dared to threaten our wonder rocket.

"Yes," added Dornberger, "and they used an engineer's intelligence to defeat a saboteur's wickedness. After the first two failures, we all knew that a success was absolutely imperative this time. So we checked and double-checked everything. Of course, we inspected each part as it was installed." He held his hand up before his face and then turned it slowly in a pantomime of an engineer inspecting the vane.

"The four steering vanes all appeared perfectly fine," he continued, "and each weighed exactly the designed amount. But one of our engineers tried each one on the test activator and found that one behaved slightly differently from the others. It weighed the same, but its center of gravity and thus its rotational inertia were slightly different. Fearing a flaw in the graphite, we had it X-rayed and discovered strange voids and dense spots, which caused us to dissemble it. Someone had drilled holes inside and inserted lead weights to fool us. But our brilliant engineers were not fooled!"

The conversation continued for a minute, and I gradually realized they were not holding me responsible for the sabotage but merely complaining about the uselessness of SS interference and boasting about their own men.

When Dornberger reached for his binoculars again, I had to force my fingers to relax and let go. He gazed through them and said, "So, our rocket is now eighty-five kilometers above the Earth, soaring through the lower reaches of outer space. In a few years, we ourselves can fly on larger vehicles of the same kind, in the weightlessness of Earth's orbit. Ah, but even today, with this perfect launch, a great weight has fallen from my shoulders. Come with me!"

We three hastened down the stairs that led from the roof of the building, and piled into a staff car, with Dornberger himself at the wheel. Everybody at Peenemünde seemed to be milling about, cheering or talking excitedly. We careened through the crowd until we found Wernher, and he rode with us to the launch area. The technicians in charge had opened the blockhouse doors, and some of them were gazing upward for a last glimpse of the rocket.

Here again we could hear the Doppler tone. It changed abruptly but continued, telling us that the vehicle had survived re-entry into the atmosphere. Then it ceased. A technician shouted into his microphone, "Impact! We estimate a range of nearly 190 kilometers. A perfect flight into the North Sea!"

Dornberger stood on top of the staff car, waved his arms for attention, and proclaimed, "Today the spaceship was born!"

* * *

I began work on plan P and plan Q immediately, more subtle forms of sabotage. In the first of these twin schemes, I interested my SS boss Hans Kammler in taking over Peenemünde. In the second scheme I tried to arrange a takeover by Gerhard Degenkolb of the munitions ministry.

All across Germany, top Nazis were building their own private industrial empires, often ruining successful businesses by their heavy-handed tactics. Either Kammler or Degenkolb could likewise ruin Peenemünde and thus prevent Wernher from proving the value of the long-range rocket in combat. Unfortunately, Dornberger enjoyed the patronage of Albert Speer, Hitler's architect and chief of the munitions ministry, so the most I could accomplish was to slow the A-4's progress a little.

Despite my best efforts, nine months after the first successful A-4 flight, Dornberger was able to arrange a personal meeting with Hitler. He hoped to convince the Führer to give the A-4 highest priority for manufacture and deployment. Knowing this would happen, I had worked diligently to prepare a propaganda film showing dramatic engine tests and the successful October flight. Wernher selected me to accompany him and Dornberger to serve as projectionist, so I then had the opportunity to attempt plan R, the most questionable plan of the entire list.

The Clarkite leadership had debated plan R back and forth, over the months leading up to my mission, unable to resolve the terrible issue of how much its success might change history, for good or evil.

We figured that it would be easy for me to smuggle one of the sarin capsules into the meeting with Hitler, because a tiny plastic sphere in an ordinary silver pillbox would not look like a weapon to the unimaginative guards. If I used it, I would certainly die along with Wernher, Dornberger, and Hitler. But then what would happen?

The only Nazi leader who really seemed ready to take command after Hitler's premature death was Heinrich Himmler, who both had his private army, the SS, and was a master of guile and organization. But an investigation of the death scene would quickly reveal that the assassination was accomplished with nerve gas. By 1943, the Germans had not only developed the first nerve gas, tabun, but had manufactured huge supplies of the stuff. German scientists were about to develop sarin itself, which is more effective than tabun but not at all different in its symptoms. As soon as they understood how Hitler was killed, they would realize that the lethal capsule was brought into the room by me, a member of Himmler's staff, and thus accuse the SS chief of having masterminded the assassination himself.

That would put the army in control of the Third Reich, possibly with a figurehead like Speer, who was simply a highly competent follower, or propaganda minister Goebbels, who was something of a weakling despite his public speaking skills. If the army held the real power, it could turn the tide of the war and prevent Germany's defeat.

One of the greatest mysteries of history is the fact that the Third Reich possessed a real wonder-weapon in its tabun nerve gas but inexplicably failed to use it. One theory suggests that Hitler's own psychiatric history prevented him from resorting to gas warfare. In the First World War, he was a corporal on the western front and was hospitalized for temporary paralysis and blindness that followed a gas attack. Scholars long debated whether he had really been injured or if his terror had caused him to adopt hysterical psychiatric symptoms that mimicked the physical inability to move or see. We will never know the truth, but possibly Hitler had such an irrational fear of gas warfare that he could never contemplate it.

But with Hitler gone, the army could unleash the wonder-weapon tabun and lesser weapons like chlorine and mustard gas against the allies in Italy, at the same time proposing peace with Britain and the United States. Threatened with the slaughter of its population by tabun-filled rockets and chlorine bombs dropped from aircraft, Britain would be forced to declare its neutrality. Then the German army could turn the full force of its gas warfare against the Soviet Union, sweeping again eastward until linking up in Siberia with its Japanese allies. Thus, the result of killing Hitler in July 1943 could be German victory.

German victory, especially if tabun-loaded A-4 rockets turned the tide of war against the British, could lead to conquest of the world followed by conquest of the solar system. Wernher might receive the support needed for expeditions to Mars, and an aggressive conquest-oriented regime might follow his initial successes with vigorous colonization. Even some of those who found this argument convincing were deeply troubled that our humanitarian goal might be accomplished by an utterly evil regime.

In the heated debate over plan R, some Clarkites had wondered if killing Hitler in a way that would discredit Himmler might save millions of people who were otherwise destined to die in the concentration camps. But the scenario suggested great loss of life on the battlefields as the German army, bloodied and angry, crushed the Soviets. The Clarkite leadership voted to avoid taking such great risks, and it recommended against plan R.

Ultimately, though, the decision would be up to me. Whatever we voted in the twenty-fifth century, I would be alone in the twentieth, with the resources to kill Hitler if I personally decided to go through with it. When Wernher, Dornberger, and I boarded the Heinkel 111 aircraft to fly to Hitler's headquarters, my little silver box with a single sarin capsule was in my handbag.

As we flew through the fog, my thoughts wandered in circles. Would I do it, or would I not? After we landed, a staff car took us to the Führer's headquarters, where we were escorted to the projection room and told to set up our presentation. We expected Hitler at five o'clock, but the three of us waited alone for many minutes after that hour. The door burst open, and Hitler entered in the company of Speer, Field Marshal Keitel, and other army generals.

I had seen the Führer only from a distance, at speeches and a few Berlin cocktail parties, and it was a shock he when he strode over to me at the projector, took my hand in his, and said, "My dear, I hope you have an interesting film for us today." I believe it was his habit to put on gallant airs with ladies and to hide his real character. But instead of evil, I saw in him great fatigue.

He stood hunched over, his shoulders bent and covered with an immense black cape. Beneath, I could see black trousers and a gray tunic, which contrasted strangely with his pale, care-lined face. In the newsreels, he appeared to have jet-black hair, but in reality his hair was an ordinary brown color. I could not summon up feelings of either disgust or respect but forced myself to bow my head slightly and say, "Welcome, my Führer, we do indeed have an interesting film for you."

Hitler took a seat near the front, between Speer and Keitel, and Wernher began the lecture part of our presentation, explaining the design of the A-4 using models and diagrams. I stood behind the table that held the projector, with my purse hanging from its strap from my shoulder. Then Wernher concluded his talk and motioned for the film to begin.

I turned on the projector, and Dornberger switched off the room lights. Before all eyes could adjust to the sudden darkness, I reached into my purse, opened the little silver box, and placed the sarin capsule between my teeth. All I had to do was bite firmly, then spit out the contents of the capsule, and everyone in the room would die in a matter of seconds.

This was a proud moment for Wernher. I knew he was not a Nazi, but after all Hitler was the leader of his nation, and this was the opportunity to sell one of the most powerful men in the world on the potential of liquid-fuel rockets. The film showed every aspect of A-4 operations, including testing, transport on a specially designed truck, and finally the October launch.

I waited, with my teeth lightly clamped on the nerve gas capsule, feeling quite ready to sacrifice my own life to achieve my mission. I had no trouble convincing myself that Hitler and the other Nazis deserved to die, but what of Dornberger and Wernher?

I loved Wernher, of course, but I loved his dream even more. If I allowed him to live, he would experience success after success, right up through the Apollo moon missions. But then spaceflight would stall, and as he slowly and painfully died of the illness that was destined to claim him, he would realize that his dream was failing right at the threshold of success.

I told myself I would bite the capsule at the end of the film, giving myself one last opportunity to contemplate the dream of spaceflight, while listening to Wernher's voice and watching the beautiful pictures of the A-4. I had seen the film several times, but now it was affecting me deeply. As the final pictures displayed the glorious October launch, my eyes filled with tears. I could hardly read the proud words that filled the final frame, "We made it after all!"

Now was the time! I tried to close my jaw, but I could not seem to move. I could hear Hitler congratulating Wernher, although his voice was hardly more than a whisper: "I thank you. Why was it I could not believe in the success of your work. If we had these rockets in 1939 we should never have had this war. Europe and the world will be too small from now on to contain a war. With such weapons humanity will be unable to endure it." I could hear Wernher, with a clear, excited voice explaining what needed to be done to mass-produce the A-4. I knew I must end this, but I stood frozen, my mind unable to break through a wall of emotion to take even the slightest action.

Then I realized that the men had finished their discussion of the model and diagrams, and Hitler had taken Wernher privately into another room. Moving very slowly, I rewound the film, placed it in its canister, and walked down the long hall from the projection room, stopping in a bathroom to spit the unused sarin capsule down the toilet.

* * *

The time had come to put my abilities as a violinist to use. There were many fine amateur musicians among the Peenemünde scientists, and they occasionally held musical evenings at which soloists, duets, or hastily assembled little orchestras would play. I suggested that it would be essential to sustain morale during the difficult period of final development of the A-4, when the scientists were focusing closely on nagging details of their work.

So I organized a string quartet, with regular open rehearsals, which any of the top men could attend. This gave me constant opportunities to talk with any of them and prepare for the desperate plan that would have to be carried out on the night of August 17, 1943.

Although there were two or three other fine cellists, no one thought it strange when I recruited Wernher to play the cello in the quartet. His style was an excellent blend of precise classical technique and romantic intuition of the musical meaning of the scores. As the organizer and a fine musician in my own right, I myself played first violin. Rudolf Hermann, who was largely responsible for the aerodynamic aspects of the A-4, played second violin. And Gerhard Reisig, who had built the electronic laboratory, completed the quartet on viola.

As I had hoped, the audience for the open rehearsal I scheduled for that fateful night included Walter Thiel, responsible for developing the rocket engine of the A-4. While we played, I felt great sadness to see the tranquil intelligence on Thiel's face, as he listened to Beethoven's music, because I knew he had but minutes to live.

Unfortunately, Wernher's accursed girlfriend, Hanna, was there also, dressed in a striking dark-blue outfit with her first-class Iron Cross pinned to the coat. She was chatting with a group of admiring men about her new plan for cheap suicide planes in which German men would give their lives like Norse warriors of old, smashing directly into allied bombers. I desperately wished she would disappear, because my plan required me to coax Wernher into going wherever Thiel wanted to go after the recital. I knew Thiel was going to die, and we would die as well if we followed him wherever he was destined to go.

After the last chord of the final movement died away, and the small audience had finished its enthusiastic applause, Wernher and I chatted while placing our instruments in their cases. "I would like to speak with Hanna," I said, "and also to let her hear some exciting new ideas from Thiel. Why don't the four of us take a brief walk? The moon is full, so you can spend the time gazing up at your destination."

"Fine," Wernher replied. "But the moon is just a stepping stone on the way to Mars and beyond. My final destination is much higher."

We four walked slowly across the athletic field, under the light of the moon and some light from the windows of several of the buildings. "Walter," I said, "could you tell Hanna the ideas about a new method of space propulsion you were sharing with me earlier this week?"

"Certainly," Thiel began. "As you know, Wernher and Dr. Oberth have been working to develop the A-4 into a winged, piloted rocket craft, the A-9, which could be boosted into space by two larger rocket stages, the A-10 and A-11. A fleet of such vehicles could establish an orbiting space station where we would assemble the deep space craft required for expeditions to the moon and Mars. This is extremely important, and we believe that a lunar landing could be achieved before 1950, and perhaps a Mars landing by 1955.

"But voyages of exploration are only the beginning, and to colonize Mars we are going to need something far more powerful and efficient than the liquid-fuel engines we have recently perfected. I believe we will need atomic power, and I have recently spoken with Dr. Werner Heisenberg about the general outlines of how such an engine might work."

Taking Thiel's arm, I interrupted, "Heisenberg spells his first name in the plebian fashion, without an *h*, and of course he is not *von* Heisenberg or anywhere else! I understand he has made very little progress in developing atomic weapons for Germany, so at best he would be the assistant of Dr. von Braun, if the two ever worked together." I winked at Hanna and touched her arm, as if praising her for selecting such a marvelous boyfriend. My real aim of course was merely to keep her and Wernher distracted so they would be happy to walk in whatever direction Thiel chose to go.

"Yes," Thiel replied, "but let me explain the interesting engineering idea we came up with. Using uranium or one of the other radioactive elements, we could build a reactor chamber and boost it by means of ordinary liquid-fuel rockets into orbit. Then we insert a number of graphite-clad rods of uranium metal into the chamber, until the flux of neutrons initiates a chain reaction, generating great heat. We pump hydrogen gas into the chamber, and it passes between the rods, becoming very hot and expanding, until it exits very rapidly from a nozzle, generating a rocket thrust. Heisenberg and I calculate that this could be vastly more effective than a chemical reaction, propelling even a very large space vehicle to Mars."

"Well, Thiel," Wernher said good-naturedly, "that is a plausible idea, but aren't you jumping the gun a bit. We have just begun to get alcohol and oxygen working together in our engines. Hydrogen and uranium may not make friends quite so easily. Old-fashioned chemistry can still take us quite a good distance."

I stumbled, and only then noticed that the lights in the buildings were no longer visible. Thiel helped steady me and commented, "I can't believe everybody has gone to bed so quickly, so there must be a blackout in effect. Perhaps a warning has been received that another air raid is on its way to Berlin. Americans by day, and British by night; their engineers must have figured they can get by with half as many beds using that schedule!"

To the north, out over the sea, we saw flashes of light, and then we heard the distant concussion of antiaircraft fire. Wernher paused and suggested, "Walter, you and I should get to our air raid stations. We have never suffered an attack here, but we should set a good example for the rest of the men."

"Yes," Thiel replied anxiously, "but first I need to run by the static engine test stand we were using earlier today and make sure the propellant has entirely drained from the lines. Even a small shock, perhaps from a stray antiaircraft shell, could wreak havoc if there is still alcohol in the line leading back to the storage tank."

That must be it! We had no idea where Thiel had been in Peenemünde, when he was killed by the massive raid of 600 Royal Air Force bombers that was destined to come that night, but the engine test stand would be a very likely place. So I exclaimed, "We must all go with Walter, because together we can check the lines more quickly, and then get both men back to their stations. Walter, show me the way!"

Hand-in-hand, Thiel and I ran quickly forward, just barely able to see our way in the light of the full moon. I didn't look back, but I felt sure that Wernher and Hanna must have been caught up in our enthusiasm or at least would run after us to argue.

Just then there was an explosion a few hundred meters away, and then another. Almost instantly, the air was filled with smoke and dust, illuminated by reddish fires and the sweeping beams of searchlights. Flame erupted from the defense guns, detonating bombs, and the ignited tanks of gasoline and rocket fuel. Peenemünde was under full-force attack!

"Back!" shouted Wernher. "We must get to the shelter!"

I was ready to die but only if Wernher would die with me. Desperately I pretended to stumble and fell heavily against Hanna, slamming her to the ground. Thiel shouted to Wernher, "Take the women to the shelter, and I will cut the alcohol lines to the test stands!"

Wernher lifted me in his arms, and Hanna scrambled to her feet and began running. I reached for my purse to get the sarin capsule, but it was gone, knocked away when I had collided with Hanna. For an instant I looked frantically for the purse in the bushes. The direction Thiel had gone was a wall of flame, and I knew he had met his fate. Seized by panic, I ran with Wernher toward the shelter.

The noise was overwhelming! I felt the concussions with my whole body, rather than hearing them with my ears. Thermite incendiaries hissed as they spread flame across the trees and houses, and ordinary bombs whistled as they fell. The white glare of phosphorus dazzled my eyes, and yellow flames billowed from the lumber yard and gasoline station. Then amazingly I found myself in the darkness of a concrete shelter, huddled with Hanna and a number of the scientists.

Days later I learned that 735 people had been killed in the raid, including Thiel, exactly the number recorded in my digital library. My attempt to add three to that total had failed, and a bone-deep sense of resignation told me that I would never again try to kill Wernher von Braun.

* * *

A year later, the first wave of A-4s was launched against London and Paris, each carrying a ton of high explosive. Propaganda minister Goebbels had renamed the rocket the V-2, "vengeance weapon number two," and now it paid the British back for their raid on Peenemünde. In response to the air raid, the munitions ministry had built an underground factory for the rockets in central Germany at Nordhausen in the Harz Mountains. But we remained at Peenemünde, working on an antiaircraft rocket called the Waterfall, which I knew gave Wernher the opportunity to develop the storable propellants that would be required for a moon landing.

By late January 1945, it was clear to everyone that Germany was close to collapse, and the Red Army was advancing on Peenemünde from the east. Dornberger had left for other duties, and Wernher received several contradictory orders from the various leaders and government agencies who felt that the rocket base belonged to them. One set of orders told us to hold the fort, fighting as ordinary foot soldiers to defend Germany from the Soviet advance. Another set of orders called for all available units to converge in the mythical National Redoubt in the Black Forest, where Hitler hoped German armies could hold out until the Soviets and western armies began battling each other, facilitating a glorious resurrection of Nazism.

Wernher decided to send all the scientists and engineers down to Nordhausen, which was near the National Redoubt. He explained to me, "The collapse of Germany is tragic, but in the history of spaceflight it will be nothing more than a minor setback. Germany will not be able to afford rocket research for many years, so we must find a new patron. The French are a broken nation, and Britain is near bankruptcy from the cost of the war. The Soviet Union is rising toward great power, but the Reds would merely crush rocket knowledge out of us like we squeeze an orange to get the juice, so I do not think we would survive their patronage. That leaves the Americans, who are both rich and naive, so I have decided to surrender to their forces, which are moving toward Nordhausen."

We would have to cross much of Germany, and there was the danger that some petty Nazi chieftain would seize us as his personal hostages, to trade with one of the advancing armies for his life. Therefore we cooked up an official-looking operation, the Project for Special Disposition, and printed red-and-white signs carrying this legend on all our vehicles. Five thousand Peenemünde employees with their families set out like Moses and his people from Egypt, by car and truck and train. Dieter Huzel, a disciple of poor, dead Thiel, buried several tons of documents, test equipment, and data in a central German mine, which he sealed with explosives, so we had some bargaining chips with the Americans.

In March, the time came for Wernher himself to make the trek southward. He and I had spent the day checking inventories on the last convoy of trucks, and it was nearly midnight when he and I were ready to leave.

The driver of Wernher's staff car was as exhausted as we were, but he saluted smartly as the two of us piled into the back seat. If I had not been there, Wernher would probably have sat up front with him. He was an experienced driver, who had made the trip several times in the past month, but the night was very dark and all Germany was in blackout because of the constant air attacks. The headlights of a speeding car on the autobahn would be an invitation for one of the British or American night fighters to strafe, so we kept ours off.

Quickly, Wernher fell asleep, and I drowsily leaned against him, overwhelmed by fatigue and a private sense of failure. Only after a long while did a stray thought wander into my mind. Perhaps this was the moment in time that would have been plan V, if I had still been following plans. At the very end of the war, when Wernher was speeding toward his surrender to the Americans, he would be in a car wreck that broke his left shoulder and arm. He had nearly been killed, and was still recovering from serious injuries when he arrived in the United States with over a hundred of his engineers to build the US missile program.

In a strange mood of anticipation and terror, I watched the shadows zoom past the windows of the car, expecting at any moment to be killed. It was so dark I could hardly see the driver's face in the mirror, but I thought his eyes were closing. My own consciousness drifted on the edge of dreams, and my senses became gradually duller and duller. Then, when I was in the drugged mood of half sleep, the driver missed a curve and lost control of the vehicle.

Instantly, we were shaken from our daze as the car thumped over the curb and began to tumble down the embankment. I screamed as Wernher fell heavily against me, and then an instant later I was flying over him. We were both thrown forward against the back of the front seat as the car collided with a stand of trees. Then suddenly everything was quiet, and we gasped to catch our breaths. Wernher was able to force one of the doors open, and we climbed out. Neither of us was badly hurt, but when we checked the driver, we discovered that he was dead.

A few minutes later, we were picked up by Huzel, who was on his way back from his mission to cache A-4 designs, and he carefully drove us the remaining hundred kilometers to Nordhausen. I stayed with Wernher until the Americans came, and because his English was poor I translated for him at their first meeting. At the first good opportunity, with $50,000 sewn into my overcoat lining, I vanished.

* * *

For a few months after I abandoned my mission and left Wernher, I threw myself into a wild life of hedonistic pleasure. I was not conscious of regret, at least not at first, and my primary feeling was relief. The stresses that I had lived under for fifteen years were over. I had carefully created an alternate identity for myself as a Swiss heiress who had established permanent residency in the United States.

After a few weeks of basking in the southern California sun and coming to the realization that Hollywood was really not very interesting, I moved to New York and took an apartment in Greenwich Village, where I thought I would enjoy the artists. Manhattan was fully established as the capital of the art world, having replaced Paris even before the war, and everything exciting in the universe seemed to be happening within walking distance of my home.

It was at a chamber music concert that my mood changed. I don't remember the name of the man who escorted me, but he was a minor painter who also dabbled in avant garde music. In the last movement of a Bartok quartet, unexpectedly my eyes filled with tears. My escort noticed and remarked that Bartok was really a victim of capitalism, having died in New York two years earlier of an illness that would have easily been cured by a little money. I commented that more likely the war was to blame, because surely Bartok's music was fashionable enough to have earned him a decent living if everyone had not been distracted by the conflict.

I left the concert at the intermission and returned home alone. Sitting in the belvedere, gazing over the brilliant lights of the city, I meditated while slowly sipping sherry. I wondered how many artists had been killed by the war, and pulled my digital library from its place at the back of a drawer to check. But the battery had run down, so I left it on the window sill where the morning sun would recharge it, and went to bed.

Over a cup of espresso, the next day, I searched the digital library's biographical database for artists, authors, and composers who had died in the years 1939 through 1945, aged less than 50. There were quite a few, and I scanned their life stories. It was afternoon before I came to the last, Anton Webern. He had been a disciple of the 12-tone school of musical composition and was famous for the mathematical purity of his brief works, in which every note was like a little polished diamond. In the confusion of the early hours of the American occupation of Vienna, he had failed to heed a soldier's command to halt and was shot dead.

A light rain was falling, and despite the chill I walked slowly to a cafe a few blocks away for a late lunch, and the sun had already dried the streets before I returned. I did a second search of the database, looking for artists who would die prematurely over the next decade, and again I found a long list.

One of the stories startled me. In just two weeks—on June 26, 1948, to be exact—the pioneer abstract expressionist painter Arshile Gorky would be riding in his agent's car, when it would have an accident. Gorky's painting arm would be paralyzed, leading to his suicide a month later. I thought back to the accident with Wernher, when my mere presence in the car had prevented him from breaking his arm.

At that moment I began my second mission. No longer would I try to kill anybody. Rather, I would intervene in the lives of artists, writers, composers, or other creative people just before my digital library said they were going to die prematurely. It was a simple matter to hire a man to disable the agent's car on the morning of June 26, and Gorky survived into the full flowering of his art, never knowing that I had saved his life.

While not fabulously wealthy, I lived comfortably on my investments, with enough left over to intervene indirectly in a half dozen lives each year. At the same time, I inserted myself far less carefully into the intimate lives of a few individual artists with self-destructive personalities. I could not count on a brief intervention to save them, because they tended to be chronic alcoholics or depressives or both. My tender attentions prolonged the lives of the poet Dylan Thomas, the painter Jackson Pollock, the playwright Brendan Behan, and the opera singer Jussi Bjoerling. The emotional strain was sometimes too much for me, and as the years passed I more and more often fell into periods of lethargy and even despair.

In 1961, I broke my rule against saving anybody over 50 and averted the suicide of Ernest Hemingway. He wrote one more novel, some say his greatest, before dying of natural causes. The next year, I hired a man to disable the car of comedian Ernie Kovacs so that he would not have his fatal crash. My agent botched the job and the accident did occur, but Kovacs was merely injured, and during his long recovery he became much more serious about his writing and evolved into a brilliant political satirist.

I never saw Wernher during all those years, but I did follow his career. His marriage was not in the papers, but I had set my digital library to give me a historical summary every week, and early in 1947 it reminded me that the American Army secretly flew him back to Germany to marry his cousin, Maria von Quistorp, whom he had hardly seen since she was a baby. All the years I had sought intimacy with him, and now I suffered the private humiliation of being bested by a near stranger.

The 1952 articles in *Collier's* magazine, popularizing the fact that rockets could take men to the moon and Mars, made Wernher a celebrity. By then he was working on the *Redstone* missile, essentially an Americanized A-4 capable of carrying a nuclear warhead, but the American public did not know about that.

Every year or so, I would check the timeline in my digital library, reading the news of Wernher's life that was hidden from the world. His next project was *Jupiter*, an intermediate-range ballistic missile (IRBM). These rockets were easily capable of launching a satellite, if fixed with small upper stages, and Wernher modified a set of *Redstones* for just this purpose. Because the army wanted weapons, not spaceships, he called his satellite launcher *Jupiter-C* and said it was intended to test warhead re-entry for the *Jupiter* IRBM. That was the same trick he pulled when he tested winged A-4b rockets at Peenemünde, pretending to increase the destructive range of the missile, when he was really developing his A-9 manned orbital plane.

Early in 1956, I knew he was ready to launch, but the army figured out what he was doing and made him fill the fourth stage with sand, so it would not go into orbit "accidentally" and upstage the supposedly civilian Vanguard satellite program run by the navy.

In 1957 the Russians launched their first satellite, as I knew they would, and Vanguard was progressing poorly, so Wernher got the go-ahead. In ninety days, on his first attempt, he succeeded. Four years later, the Russians orbited the first human, and Wernher had extended *Redstone* and *Jupiter* technology to create *Saturn I*. Despite the similar name, the *Saturn V* that launched the moon missions was a thoroughly American design, but a hundred of his Germans still held key jobs in the American space program.

Late in 1968, I underwent an operation, and while recuperating I watched the television coverage of *Apollo 8*, the circumlunar flight that Wernher had described back in 1952 as the last step before an actual landing. I insisted that the doctors be frank with me. The thirty-ninth anniversary of the day I landed on Earth and met Wernher came, and I would probably not see the fortieth. Whatever the calendar said, biologically I was nearing seventy years old, and I was tired.

Over the years I had often seen my digital library's copy of the movie *2001: A Space Odyssey*, based on Arthur C. Clarke's writings, but now I had the glorious experience of attending the premieres in several nations around the world, watching it on the wide screen with audiences who were mystified almost as much by the imagined technology of the twenty-first century as by the surreal scenes or the black alien monolith.

In the middle of May 1969, one of the most prestigious galleries in New York auctioned my art collection, and I sent anonymous stipends to more than a thousand young art students around the world. I wrote a long letter to Wernher, explaining my mission to kill him and including papers documenting I had often known about important events before they occurred. It did not matter now what he thought about me or what either of us felt about the other. I had to make him believe me so he would undergo surgery to remove the apparently healthy kidney where his fatal cancer was destined to bloom in just a few short years. I had set out to kill him, but now I was giving him information that might add a decade or two to his life. No one could do the same for me.

Fatigue gripped me, and I hardly had the strength to weep. I resolved to end my life before my illness did in its inexorable course.

Suicide would be my last act of free will, if any of us truly are free. But I wanted to live just long enough to see Wernher's triumph.

* * *

On June 22, I lay on the bed in my Greenwich Village home, watching the television coverage of the landing of *Apollo 11*. Like millions of people all around the world, I heard Armstrong's voice, "Clavius Base here. The *Eagle* has landed."

Only recently had I realized that some of the details of the flight were going to be different from those recorded in my digital library, launching a month early to a different landing site. It had taken me several days of pondering to decide that I had changed the history of spaceflight, ever so slightly, merely by being in the car with Wernher when it wrecked in March 1945. Merely because I was there, not from anything I intentionally did, Wernher was spared a broken arm. He surrendered to the Americans in somewhat better condition and was able to begin his work for them slightly more quickly.

I imagined Wernher's joy, and his pride, at the success of *Apollo 11*. But I knew that spaceflight was still beginning a century too early and would soon burn itself out. My mind was spinning, partly from a lifetime's repressed emotions, and partly from the moderate overdose I had been taking of the narcotics prescribed for my pain. In a few hours, I would have no further need of them. The little silver box on the table beside me was ready for the moment the crew finished its brief walk upon the surface of the moon.

Six hours later, I heard Armstrong say his famous words, "That's one small step for a man, one giant leap for mankind." With brandy, I washed down several of the white tablets and reached for the silver pill box in which I kept the one sarin capsule I had saved for thirty-nine years.

"Magnificent desolation!" Armstrong's voice again. "But I see something else. Perhaps it's just a shadow, but I am going to take a closer look."

I placed the sarin capsule in my mouth and clamped my teeth down on it. I felt a pleasant warmth slowly building in my body, yet my muscles shivered involuntarily. The television seemed very far away, but I heard the voice of mission control. "Neil, did you get your contingency sample? Stay near the LEM, Neil, and get your contingency sample."

"Negative, Houston, I've found something." On the screen, I thought I could see something, too, beyond the lunar lander, a regular shape that did not look like a hill or a shadow. "Houston, there is an object here, not something natural, but manufactured. It's almost as big as the LEM. Dust on it, like it's been here a very long time. Buzz, can you bring the television camera over here?"

"Neil, get the contingency sample!"

"Negative, Houston, what we've found here changes everything."

I felt as if I was floating in space again, and the room was so dark, so dark. With tremendous effort I tried to see the image of the monolith on the moon, and then I was enveloped in light.

Chapter 2

Annual Meeting

Carolyn Mattick and Brad Allenby

DRAFT 4/27/2021
Script Copy. Top Secret. Do Not Reproduce or Distribute.

Smile. Be engaging and yet naive, as if you can't believe your good luck. Lay the groundwork for innocence as a response in case outside agitators manage to get a question in. Remember that our strategy is to be open and honest to the extent possible. While Security has assured us that it has identified and managed the physical attacks that were being planned, it is neither possible nor desirable to bar antitechnology activists. Remember, however, that they will be operating from scripts as well—including scripted activities—and that they will have real-time recording devices. Accordingly, they will be identified by augmented reality markers in your vision field and are not to be recognized. Security has identified two organizations that will conduct performance art disruptions; our current plan is to allow them to occur without interference but to enable local noise-cancelling technology and limit their physical extent so that the Meeting may continue.

Flash Forward: A Series of Futuristic Vignettes
Edited by Nora Savage and Anita Street

ISBN 978-981-4669-44-3 (Hardcover), 978-981-4669-45-0 (eBook)
www.panstanford.com

Welcome, ladies and gentlemen, to the fourth annual shareholder meeting of Integrated Foods™, "Purveying High-Quality Nutrition to the World since 2017." I am proud to say that we have been the fastest-growing company in history, thanks to our cost management, reliability, and unbeatable safety record. (*Shake head in humble amazement.*) We are the largest commercial operation in the food sector in 14 of the countries within which we operate, and our Chinese and Brazilian operations are spectacular examples of how those global powers have managed to consolidate their positions so rapidly. Our accomplishment is even more remarkable given the ongoing campaign by obsolete, inefficient, and environmentally dangerous competitors who can exist only by an increasingly unacceptable reliance on animal cruelty and exploitation. We make the world's food in clean industrial settings without using and abusing land, nature, and, of course, the poor animals that are still, sadly, killed in brutal ways every day. (*Face camera squarely with sad and indignant expression, as you have been practicing.*) And we can do far more—but I'll talk about our new product line a little later.

It is worth looking back briefly to see how we did it.

It all started when we opened the first Biomeat® plant outside Kansas City. We hoped to use emerging tissue engineering techniques to provide the world with a safe, affordable source of protein produced entirely within our factory, while reducing animal suffering and limiting the environmental impacts of meat production. We started by taking a small sample of muscle tissue from Greta, a beautiful Charolais cow who continues to live in the Kansas City resource garden. (*Gesture to the monitor showing a picture of Greta relaxing in her luxurious habitat.*) When stem cells from the sample were added to a bioreactor containing water, glucose made from corn starch, and a few other compounds, the result was several thousand pounds of humane and environmentally friendly meat every year. We had lofty goals back then, but the results have surpassed our wildest expectations: The number of animals slaughtered in the United States has decreased about 15% every year since 2019, and swine and bird flus are down 25% worldwide. The land that we have freed up by doing this has helped keep agricultural prices stable, even as biofuel and bioplastic production has accelerated.

This success inspired us to do more. Six months after that plant opened, with the help of a USDA grant, we embarked on a major renovation to improve the plant's environmental efficiency, with the goal of establishing a zero-waste, fully sustainable production process. We learned to think like nature, and in so doing, we realized that we could recycle all of our nitrogen for reuse. Other waste products are composted for precision application to our agricultural resource gardens. As it turns out, we failed to meet our zero-waste goal: That flagship plant now returns 50% of the water it uses to Midwestern farms—but it's clean enough to drink. (*Give the audience a wry smile. Hold up the glass of water and take a sip. Nod to reassure the audience that the water is completely safe.*) Getting to this point wasn't easy, but it was well worth the investment. Our practices have single-handedly reduced water pollution by 20% in the Mississippi River and meant that the infamous Dead Zone in the Gulf has shrunk every year we've been in production. Given the resemblance of the Kansas City plant to a natural nutrient-recycling system, we now refer to all of our plants as "cosystems."

Pause here. Let a concerned expression come over your face. Assume a more serious tone of voice.

As you may be aware, Integrated Foods™ has been the target of criticism by activists, some of whom are exercising their free speech rights in this hall right now. Many of these activists claim that our hyperefficient food production methods include monoculture cropping that may increase the vulnerability of the food system to damage by pests, diseases, and extreme weather. I want you to know that we are sensitive to all threats to human and environmental well-being, and we feel that security of the global food supply can be achieved through diversity and redundancy at the global scale. For this reason, in 2018 we opened ecosystems in Europe, Australia, Brazil, China, and Angola. Each ecosystem is designed to work with crops derived from plants traditionally grown in the region: sugar beet in Europe, sugarcane in Australia and Brazil, rice in China, and cassava in Angola. This not only enhances food system diversity but also preserves local farming traditions as we work with local growers to establish resource gardens that provide our raw inputs. (*Gesture to the monitor, which will show smiling farmers standing next to peaceful fields with healthy plants swaying in the wind.*)

At the same time, if one crop or facility should fail, the others stand ready to ramp up production and ensure that no one goes hungry. Equally importantly, we have found that nations experience an average increase in per capita income of 5% in the year following the establishment of an Integrated Foods™ ecosystem.

The next paragraph will be in all our media feeds, so make sure you practice it. It has been approved by our lawyers, so it must be delivered exactly as written. We have placed articles detailing these charges with WaPo, WSJ, *and* NY T*embargoed until the speech is given. Talking points have been distributed to executive-level staff.*

We will always appreciate honest criticism; it helps us be a better company. But I must tell you that we have discovered that these well-meaning people are being manipulated. Using US and international law enforcement and election records, which are public documents, we have tracked funding and leadership of the antimodern food movements to firms that are being challenged by our business. (*Be a little emotional here.*) We embrace fair and open competition, and we have always encouraged alternative business models for the farmers who may be affected by our expansion. But if a company has failed the poor, abused antibiotics, hurt animals, damaged nature . . . if that company's response to our better way is an underhanded and secret campaign misleading and seducing idealistic young people like those you see carrying the signs outside this hall, that's not competition; that's shameful. (*Be statesmanlike here; this will be the 20 seconds that make YouTube.*) Sadly, we have also found illegal international contributions to certain politicians; names, amounts, and evidence have been turned over to Interpol and appropriate national forces. We produce better, cheaper nutrition for the world; our competitors bribe, cheat, and even attempt to suborn the political process, trying to stop us.

And yet we continue to grow. In 2019, we let nature inspire us once again and expanded our facilities to produce fruit and vegetable concentrates. These are potent sources of bioavailable vitamins, minerals, and antioxidants, many of which have been associated with cancer remission. We obtain these substances by culturing only the most nutritious parts of plants. This allows us to deliver high-potency, naturally sourced nutrition with less waste. In fact, these concentrates far exceed the vitamin concentrations in superfoods like kale and blueberries, and we can produce them from any agricultural source of starch. This reduces the variety of crops that need to be grown and further increases agricultural efficiency. (*Gesture to the monitor, which will show a vibrant tropical rainforest.*)

Plant concentrates have been a smashing success and are currently responsible for 30% of global revenue. You can find them in our Whole Nutrition® line of shakes and 3D-printed crackers designed to meet the specific needs of your individual health profiles. All you have to do is visit one of our kiosks at participating grocery stores, where a swab from your cheek will allow us to quickly and painlessly recommend the best formula for you. Just don't eat too much unless you want to live forever! (*Wink at the audience.*)

Pause briefly and look at the audience with sincerity and empathy.

But as shareholders, you want to know what the future looks like for Integrated Foods™. I can assure you that it is very bright indeed. When we entered the food industry in 2017, it was a $4.5 trillion market dominated by processed foods. That year, cultured foods became a new category in the world food market, and it has since acquired a significant market share. Last year, cultured foods were 30% of the market, and we expect strong continued growth until saturation is reached in 2030. At that point, we anticipate that fresh foods will make up 2% of the market, primarily cultured foods will make up 48% of the market, and processed foods that may contain plant concentrates will constitute the remaining 52%. So, as you can see, even though competitors will undoubtedly emerge, Integrated Foods™ is poised to capture a large portion of a $5 trillion market. (*Wait for applause to fade.*) After 2030, growth will occur primarily through gains in per capita income spent on food, as well as population growth, which may stabilize as soon as 2050. (*Gesture toward the monitor, which will show significant future revenue growth beyond 2050.*) Please refer to Section 6 in your copy of the annual report. You'll see that, in spite of current population projections, we are preparing to maintain strong revenue growth for decades to come. We plan to maintain our advantage through strategic innovation, safe production practices, and public engagement.

In terms of strategic innovation, we have several new offerings currently being tested. I am very excited about our new Probio™

(*proh-BAHY-oh*) line of frozen meals. These contain a proprietary mix of healthy bacteria that help regulate calorie uptake and keep consumers looking fabulous. Why should anyone have to eat less of our delicious food to stay thin? Our bacteria are engineered to consume the calories people don't need and convert the rest to heat, and nontoxic compounds for excretion. (*Smile optimistically—by the time it goes to market the damage to stomach and intestinal walls will be virtually undetectable.*) The best part is that the packages are self-heating. Just peel away the plastic cover and eat! (*Act as if this is the best thing since sliced bread. This is truly a brilliant way to prevent the bacteria from being killed during the cooking process.*)

Product lines like Probio™ will help Integrated Foods™ capitalize on the ongoing process of global development and expand the size of the aggregate food market. Additionally, existing protocols will serve to ensure continued growth of the market. In particular, a healthier population with greater longevity will undoubtedly contribute more to society and consume more high-quality food in the process. Due to our sterile food production and handling practices, we can already report that human deaths due to food-borne illness have dropped 50%—and we have achieved that without using antibiotics in ways that encourage global development of resistant, harmful bacteria. We even anticipate that prion diseases such as Alzheimer's will decline in future generations as a result of our rigorous testing protocols. This translates into longer, more productive lives—and more opportunities to be Integrated Foods™ customers. For the first time in history, food really is medicine. In fact, I believe that we make the food that Mother Nature would design if she could start all over again.

The third tier of our strategic plan involves educating the public about the new state of food production. Due to our hyperefficient, environmentally beneficial production practices, we are no longer constrained by scarce land and water resources. The future is indeed bright for the planet, and there is no longer any reason to limit population growth! In fact, a healthy global economy may depend on an increasing supply of productive labor. We have therefore engaged a marketing firm to show people how satisfying life can be as part of a large family. With their help, we at Integrated Foods™ believe we can create a world full of happy, healthy, productive people! (*Gesture toward the monitor, which will show the smiling faces of children from all over the world.*)

Finally, let me note that our new CelebrityLine™ sausages, containing genetic material from famous film stars, continues to position us as the cutting-edge integrator of culture, humans, and their individual genomes, food, and pharmaceuticals. In response to overwhelming public demand, we will be expanding this product line to include designer foods for weddings that contain genetic material from both the bride and the groom. We also plan a historical personality line, when we can legally obtain the necessary starting materials. (*Assume a sympathetic expression.*) While we regret the injuries that occurred in the riots at the opening of Johnny Depp's latest film, *Jumping the Caribbean Shark,* which for the first time offered snack foods constructed from the star's own stem cells, we believe that continuing innovation is, in a real sense, in your company's DNA (*finish the segue from the serious topic of the riot by chuckling slightly here.*)

In closing, let me reiterate that both humanity and Integrated Foods™ have very bright futures. Together, we can ensure the prosperity and health of all the world's children. Thank you for your time, ladies and gentlemen. Don't forget to get your Whole Nutrition® recommendation before you leave. There are kiosks in the back of the room, and samples are available in the hall. I'll be happy to take questions now.

Smile and wait for the applause to die down. Try to take questions from the actors who we've planted in the audience—they will be identified by green tie markers in your augmented vision field—but should someone else get a question in, here are some talking points . . .

Product cost: Integrated Foods™ has a variety of product lines that appeal to people from many different income levels. In general, our focus is on quality for enhanced human health and performance—and quality has always come at a premium. We do, however, have our Complete Nutrition™ line of inexpensive products; while these are obviously not individually designed and do not include all premium ingredients, they are popular in many markets in developing countries.

Energy use: While it remains true that our ecosystems require a great deal of energy to operate, I assure you that our energy supplies will not run out and our energy outlays will never exceed 3% of operating costs. Due to brilliant foresight on the part of our finance team, we have established hedged contracts with several energy companies to maintain fuel reserves specifically for our ecosystems for a fraction of the market price. Moreover, all new ecosystems are planned with natural gas–combined heating, cooling, and electric power generation systems on-site: the excess electricity is used to operate our carbon dioxide capture technology units, so we actually reduce atmospheric carbon dioxide as we operate. A few scientists and engineers from the antimodern food movements have claimed that the carbon capture system is a net carbon emitter due to the power it consumes. They say we would be better off scaling back our generators or selling electricity to surrounding communities to offset less efficient power production. But that all seems like funny accounting to me! Anyway, older ecosystems are being retrofitted with this technology as capital flow permits. This arrangement will ensure that global food supplies will always be safe and abundant.

Climate change: We are aware that increasing temperatures associated with anthropogenic greenhouse gas emissions could impact agricultural productivity. For this reason, we are constantly researching new plant varieties that will increase starch yields under a variety of climate conditions. It is also important to remember that the Integrated Foods™ ecosystems are much less vulnerable to changes in temperature, rainfall, and soil quality than food production has been historically. And, to promote wider use of our patented carbon capture technology, we are proud to be a founding member of the Climate Management Operations Network, a group of companies that link their technologies to enable rational, ethical, and responsible management of fundamental natural cycles, including not just climate and the carbon cycle but the nitrogen, phosphorous, hydrologic, and sulfur cycles as well.

Consultant fees: As some of you may have noticed, the financial statements in Section 9 of the annual report indicate that a sizeable outlay was made to consultants last year. To ensure the continued success of Integrated Foods™, we employ a number of consultants and advisers to help us project market trends and maintain a competitive advantage. Occasionally these activities extend to political engagement, including supporting candidates and sponsoring legislation that would benefit our shareholders. Lately, to ensure the continued growth of our customer base, we have hired firms to educate various government officials on the value of continued population growth, specifically to ensure continued global economic prosperity post-2050.

Whole Nutrition® kiosk terms of use: Some consumers may have noticed that the Whole Nutrition® kiosks ask consumers to check a box indicating that they have read and agree to the terms of use when they submit their samples. These terms are available on our website, but I can assure you it is merely a legal formality. It prevents Integrated Foods™ from being sued in the event that someone accidently slips and falls while using a kiosk. Conspiracy theorists have made outlandish accusations that we are compiling a sizable database of human genetic information. Some even claim that we mine this database for sequences that could be spliced into our animal cell lines to make them grow faster or taste better or whatnot. They say that we are profiting from genetic material that does not belong to us and that we are selling chimeric meat—meat from genetic hybrids of multiple species. Let me assure you once again, ladies and gentlemen, that we at Integrated Foods™ have only the best interests of our shareholders and our customers in mind when we develop new products.

Chapter 3

California Dreamin'

Brad Allenby

Overture

I had the package funeral. It was in a run-down wedding chapel five blocks off Las Vegas Strip, behind a junkyard littered with pink Caddys. I was squeezed in between a couple of teenagers who barely stopped pawing each other long enough to say "I do," and two drunk oldsters reeking of Formica. You could tell it was a funeral because the jefe, obviously channeling Lady Gaga or some such, substituted a snazzy black and gold cape for the white, gold, and pink wedding version. Otherwise, it was the same: the mandatory witness, twirling his hands in some possessed drug dance with his demons, in the fifth row; floating cherubs, bloated and chipped and discolored, smiling on my demise in fat-cheeked dimpled glee; three live streaming cameras around the room, one of which actually worked. One of my kids had promised to watch it live, but her work schedule changed,

Flash Forward: A Series of Futuristic Vignettes
Edited by Nora Savage and Anita Street

ISBN 978-981-4669-44-3 (Hardcover), 978-981-4669-45-0 (eBook)
www.panstanford.com

so she was waiting tables instead. S' ok … it's stored for six hours before deletion.

Liberace up front waxed poetical about my life, although he had no notes, and we had little chance to talk beforehand. But he knew me well. Apparently, I had been quite a power at the CIA, and—in obvious violation of my current state—a real social star, popular with the ladies … who knew? My five minutes was next: I tossed my notes, agreed with capeboy since his version of me was a whole lot better than mine, and said good-bye in the direction of what I hoped was the working camera. And done: for the next six hours, as real as I would get.

To be fair, it isn't the same as it used to be, back when people were really dead. Everyone today opts for the wetware special: trade ownership of your genetic and proteomic information for radical life extension. Sure, you end up leasing the chassis, but in return the chassis keeps running. Not an unreasonable deal. But make sure you read the fine print about memory management.

And … boring. Really, really, boring. The psychoactive materials and entertainment sectors have boomed, of course, and you have to appreciate the technical advances. Waves of sheer information long ago eclipsed individual processing. Anything rational comes from networks now. And none of us understands; we just experience. All the old dystopian hysteria of soma, and religion as opium … how little they knew. Integrate your drugs and visions directly into your central nervous system, rock your networks, and reality doesn't have a chance. Ask that guy in the fifth row.

Ask me. I dunno my age, and I lost my past a long time ago, dropping like chunks of rotting flesh from a hard done undead. Don't even remember when my computer woke me up. It was the Net calling. Again. Usual terms. Flippin' Net sucks human brains like a bad zombie movie. Puts 'em in some sort of *n*-dimensional space and uses the raw computational power to feed its exponential AI liftoff. That old black magic cultrock *The Matrix* had the right direction, but way wrong function. Humans are lousy energy sources. They are great wetware chips. The AI wave now sweeping across space and time is Powered By Brain. A lot of brains, actually. Wired in parallel. Everywhere. I think that's why even now, even when meaning isn't accessible anymore, there's still sex. Hormones make little brains.

And the Net's always looking for more. To the pathetic point it kept poking ... even me. And this time I said yes. I felt the Net rub its grubby little paws together. Another chip. And you know what? I'm not sure if the funeral was part of the package, or even real. But now, I don't care.

Buffo: Death of the Hero

I knew the car. It was a bright red 1966 389 Tri-Power Goat convertible with a 4-speed and serious rubber. It was doing 110 just sitting in my driveway. It would do 110 all the way to school, even if the speedo and the cops said 25. It was born doing 110. It was growling quietly to itself, or maybe that was just the garbage truck down the way. She—and oh, lord, she was a she—had curves that made anything and anyone else I had ever known look like a failed wetware experiment gone to putrescence. My new rule, right then, was always date your car. Love your car. Adore her. Until death. Her chromed fenders are more truth than you'll find in your entire school. And they're real, unlike your adolescent holding pen.

Especially if it comes fully equipped. I didn't know the girl. She wasn't in my class; I definitely wasn't in hers. But when she looked at me, I could feel organic acid, intuit biochemistry. Carbon bonded, double bonded. Everything about her coupled into a neuron somewhere. I could feel networks rearrange. She didn't look like a cheerleader. She looked like a Beach Boys song. Her hair—blonde, very blonde—drifted in the breeze like an extended lead guitar riff cut loose to dream. Her smile was the caress of endless nights on Southern California beaches, and I couldn't tell where the sunlight stopped and her figure began. Suddenly adolescence made sense.

Or, rather, didn't. Never look gifted Detroit Iron in the mouth, but ... that part of my gig was long gone, blessedly repressed, a sad reality show notable mainly for its casual cruelty and my inability to nail a scene. Any scene. Undoubtedly, my fellow thespians hadn't seen it that way, prancing their way in self-absorbed narcissism across the volatile stages of teenhood and high school—but then, they weren't the sludge lining the silver cloud of teenage social fantasy. Votech players speaking the language of carboration, skank to hand. Topsiders and khaki, the *Ancien Regime* gone to seed. Self-selecting constellations of cool, with globular clusters drawn through the invisible gravitational attractions of hyperaware insecurities. Islands of female floating in inchoate hormonal seas. And then ... the few, the downtrodden, the geeks. The problem with reality is that when you're voted off the island, you got no place else to go, and everyone in the game knows it. Voting you off just makes you fair game. And I was an obvious choice.

And there she sat, radiating red and silver; and there she sat, mythic California made flesh. There was an issue here. Serious cognitive dissonance. Or something dissonance; cognition as unnecessary complication.

That's how she played it. Music, so soft it ruffled at the edges into orange blossoms and morning surf. She giggled, realizing that I hadn't been tracking anything she'd said, and my lovely car insolently stretched, amused, steel on steel. Words coalesced out of a light morning fog.

"So ... you coming, or not?"

"Umm ... "

"Don't have to. But she's got a full tank. She always does."

" ... to school?" I wanted to add "again?" but there's this fear when all three cherries show up on the spinning wheels that if you breathe, the universe starts again. And I couldn't bear thinking about that. Better not to breathe ... or to let four barrels of sculpted metal do it for you. Far as that goes, it also wasn't clear that I could bear thinking at all. Vastly overrated.

She knew. And so did Suzy Q. She was Suzy Q. I could have spent years just in love with the fenders. Straight pipes and chrome exhaust. She had to be Suzy Q. For when she was sultry. Suzy when we were less formal, out running together. When she was being formal, red hips and black paws streaking the roads with burned rubber elan, she was Suzanna. And I knew, even without sliding onto the black leather bucket seats and holding that shift, we would not be going to school. If you have four forward speeds, you don't need reverse. School was out.

"Route 66 ... "

There were words, and maps, and places, and plans, but they smashed against the reality of endless asphalt the way bugs exploded in pieces against the windshield. I couldn't tell if it was Suzy or California ... could it really be Suzy? And then ... I didn't care. She was someone that I had always been in love with, in adoration with, and always would be—but I really didn't know much. I didn't know how to know much. That was obvious even to me. Suzy Q purred. California infused me. Actually, not much was obvious.

You may talk of your cylinders, of your cam, of your valves marching in serried rows down city streets gated by massive clutch plates, but oh, how little you know. You may talk of fluids and sex, and biogeochemical hormonal signaling systems, and plastic surgery of all modifiable surfaces, but how little, how little you know. As I opened Suzy's door the Rubicon yawned, and I crossed it. I did not ask, I did not cling to my present, and you who throw the banalities of today around as if they validated your soporific reality ... you know not of what you speak. You manufacture a caricature and claim it is meaning. Not so ... not so.

The white coats of science ooze journal papers speaking of the coupling of mechanical and biological systems and computer brain interfaces. They have never endlessly, endlessly been dissociated into California. They never met Suzy, never felt her black leather embrace your mind as you settle back, never felt California blonde blowing over you like sunrise over a magic sea, the burble of Detroit Iron pumping your own blood.

Even so, even so ... how do you get Malibu and no Watts? The '60s and no Vietnam? From afar, I watched the questions tick beneath the surface, firing like a reptilian brain. Oh, I didn't want answers. Cold, logical answers, dismantling the perfect balance of power and beauty, the combination of shes that already defined me. Road trip indeed. But the ticking kept on, never conscious, but never allowing me to dissolve into the music and the surf, leaving me to hear the words that slowly, clumsily, occasionally fell from their throbbing eternal patterns.

And so we left. I'm not sure where we left from. Does it matter? I thought we'd go cinematic, sailing off a cliff somewhere in the painted Southwest, a flash of destructive performance art seen by no one arcing across the sky for eternity. How else do you want to die? How else does it even make sense? Only in the arms of Suzy, with California to carry us home. But we don't. We drive the red sandstone–sculpted planetary bones, the coastal highways with dollhouse piled on dollhouse, desert stars above the flare of Suzy's headlights. The roadhouse is always open; the cops always on some other road. Sometimes my arm is around California; sometimes the music and the waves are around me. I still don't know what she looks like. I don't want to know. I don't need to know. And always there is Suzy.

Curtain Call

I suppose some time has passed. Maybe. Hard to tell, really. But I have been permitted to understand some things. I know that I am integrated into a massive parallel processing system from which emerges what the hoi polloi call reality. They claim to understand it, because they watch screens that tell them they understand it. But each of them is only a little fragment of color, a pixel, and it is only when the networks are constructed that the image appears. The networks have knowledge that the pixels knoweth not, and it is only a cute but illusory credulity that keeps the masses tractable. The Net is not many things, but it is a damn fine shrink, and knows exactly how to drip the dope into its constituent, sparkling nodes. I know I don't know, but I also know I don't need to know, either. On beyond pixel, that's me.

I think I am being paid. Paid for staying awake, poked by the passing feel of Vietnam and Watts and the way high school really is (was?)—because otherwise I would be one with Suzy and California, one in a way you peasants can only dream of, and only then in emotions that fade to normal when you wake. Paid in nothing, really, because Suzy's singing as we blow the doors off a Caddy or a slumbering beetle, and California's effervescent haze, are not touchable. I don't want them to be. Real existence is not wetware. I don't want to live in a world where I have to touch or see a face, or hear a voice and words, to adore. I don't want to live in a world where I can't dissolve, where increasingly there will only be a watercolor wash between me and Suzy, me and California.

Funny. Used to be we were in the mind of God. So Bishop Berkeley said. Or part of it, maybe, depending on how inclusive your guru was feeling that century. Now, it's the Net … the net of all of us, and the chips and machines and interconnected brains. It thinks. We process our little stream of data. We will never know what it thinks; we can never know what it means to think. Except, as I ride the stars in a throbbing V8, and California wisps around me like all the real there ever was and ever will be, I know that it thinks us back, and maybe it always has.

Chapter 4

Intelligence on Earth?

Mike Gorman

It was a good sky—no moon, only a few wisps of cloud, and the light pollution from Charlottesville. He knew the stars only seemed permanent and unchanging because of humanity's brief time horizon and because we were in the Appalachia of the galaxy, far away from the lights and black holes of the center. He turned his binocular mount toward Sagittarius, which marked the center of the Milky Way, and fiddled with the focus. *Lights of the big city, and here we are out in the boondocks of the galaxy. Wonder if there are civilizations growing up under a sky full of stars, with neighbors? Wonder when I will laugh again.*

"Yo, Dad! Another SETI alert."

He pulled away from the binoculars, blinked. She had said he would not be a fit parent. He wanted desperately to prove her wrong.

"You can click it off. But it's getting pretty near bedtime."

Something was just out of sight, just out of hearing, pressing for his attention. *Took my Abilify this morning—no hallucinations, or Patrick will have to go to the lights of the big city with his mother.*

Flash Forward: A Series of Futuristic Vignettes
Edited by Nora Savage and Anita Street

ISBN 978-981-4669-44-3 (Hardcover), 978-981-4669-45-0 (eBook)
www.panstanford.com

"Dad, you OK?" He was right next to them, looking at the screen—no memory of getting there.

"Um, yeah."

He liked the premise of the game: exploration of new planets by multiple interstellar civilizations. Somehow the end result was always war, which the boys loved—spaceship battles and planetary annihilations. Martin would have preferred the discovery of interesting alien races—so alien humans could barely understand them.

The SETI window popped up again. Patrick sighed impatiently.

"Can you save the game? The only way for me to get rid of the alert is to read whatever—"

"But I'm not finished with my turn."

Too late, his fingers had already eased Patrick off the keyboard and clicked the link.

"Dad!"

"It's OK, I want to see, too." Martin liked Snehal

The title of the SETI message said, "WOW!" Underneath, he noticed a link to the NRAO.

Am I really seeing this? It didn't feel like a hallucination, but he wanted to be careful around Snehal. "Looks interesting, but I can check it out after—" Too late, his fingers were operating independently A powerful radio signal had been picked up by the Allen Telescope Array. The wave varied in an unusual fashion that had not yet been deciphered but did not correspond to a known signal. Its source had not been pinpointed, but it was in the direction of M55—where the WOW signal had come from.

"Dad?"

"Sorry Patrick. I need to study this."

"Dad, I need my own computer."

"I know. Your mom and I are discussing it."

"Do you think this is a signal from another civilization?" Snehal asked.

"Don't know—we have had a lot of false alarms." He did not let himself hope.

"Hey, have you guys done any homework?"

The two boys looked at each other.

"Snehal, that's the first thing your dad is going to ask about when he comes. I like having you here with Patrick. Ah, the WOW was detected, coming from the same place."

"WOW?" asked Snehal.

"A radio signal that Philip Morrison, a famous physicist, had predicted a civilization might send, corresponding to the structure of hydrogen. So it could be contact, or it could be whatever kooky even caused the first signal. Apparently this time the WOW is in the middle of a much longer signal that may have a complex pattern. Patrick are you reading this, too?" Hallucination check.

"Yeah, Dad." He did the vocal equivalent of rolling his eyes.

The phone rang. Martin jumped to it—maybe it was one of his colleagues at SETI—

"Hi, Martin. Need any help?"

"Um, hi, Grandma, I am trying to get the guys going on their project together—"

"Say no more, I am coming over."

"I'm sorry that I can't do a better job. I don't want to inconvenience—"

"Nonsense. I love seeing my grandchild. We want to look good for my daughter tomorrow."

"Thanks, ah, God bless you!" Martin felt his eyes moisten. Grandma was a great substitute for his own mother, rest her soul. What an odd expression—"rest her soul."

"Be there in a minute." She had hung up. He had wandered off again. Was she waiting for him to say goodbye?

"Grandma's coming; so is Dr. Krishnamurti—time to get to that project." The boys were already back into their game. Martin looked over their shoulders. He still had that weird feeling that he was missing something, that someone was trying to contact him. The game seemed to ease it—little spaceships jumping from system to system, exploring planets they thought were likely to be habitable by their species. There were a lot of them, and travel between was very quick. Wouldn't it be cool to be in a universe like that—

"So you've been swept up in the game, too!"

Martin stood up so quickly he almost hit Grandma.

"Oh, sorry." I was traveling between planets when I should have been here.

"Same old Martin. I didn't mean to startle you. Right boys, off the game, onto the project!"

"Just a few more . . ."

"Martin, give me a countdown from 10. At the end of it, I will pull the plug."

"10, 9, 8 . . ." Grandma was a pro—by 2, they had saved the game and were moving toward the sofa.

"No, dining room table."

"But it has stuff on it!" Patrick objected.

"Yes, and we are going to clear it off."

End result—both boys sitting opposite each other with notepaper, Grandma at one end, Martin at the other.

"Martin, if you have work to do . . ."

"No, I mean yes, I always have work, but I want to learn from the Master."

Grandma smiled. "Tell us about your project, lads. What's the first thing—"

The doorbell rang.

"It's my dad," said Snehal. "I'll get it." Martin jumped to be right behind Snehal at the door, straightening his shirt as he moved.

"Hi, Dad."

"Please come in," said Martin Nathan's finely chiseled features and alert eyes moved into the light. Martin wondered if he slept in his suit.

"So, tell me what progress you have made, son." Martin looked down.

"Patrick and I were, um, well, we saw this SETI alert."

"Oh." Nathan tilted his head to look at Martin.

"Can—can you sit for a minute. You know my wife's mother."

"Mrs. Olson." Nathan bowed.

"Oh Dr. Swami, you know you can call me Emma. I'll make some tea?"

"And you may call me Nathan. Thank you, I would love a cup."

It is amazing how much time we humans spend on these pleasantries. By simply sitting upright in the old chair and staring down at the boys across the scuffed table, Nathan made the setting seem more dignified.

"So, boys, tell me about your science project."

"We want to do it on this SETI alert," Snehal blurted. Patrick looked at him and said, "Yeah, that's our plan."

"Hmm . . ." Nathan was a Full Professor in the astronomy department, with a specialization in computational modeling of likely exoplanet scenarios to predict situations where Earth-like words might occur. Martin had a shared lectureship between astronomy and psychology; he was a PhD with a undergraduate major in astronomy and, with that hybrid background, considered himself lucky to be able to teach a few cases. Nathan had real insight into the process by which science worked and didn't work.

"So what do you know about this alert?" Nathan asked.

"Well, it came in while we were on the computer." Patrick looked hopefully at Martin, who simply looked back. Let's see him explain this one.

"And" Snehal said, "the WOW signal came back."

Nathan raised his eyebrows. "What is WOW?"

Patrick and Snehal exchanged quick glances. Snehal kept the ball. "I think it is something to do with hydrogen—"

"Right. A good place to start your research—find out everything about that signal and why anyone cared about it."

Martin's jaw almost dropped. He's going to let them do this project!

"How was it detected?" Nathan continued the grilling.

"Radio, it was radio," said Patrick.

"I hope that is enough for your teacher tomorrow, because it is time for us to get going, Snehal. Thank you for the tea, Emma, and thank you for taking the boys, Martin."

"Sure, I like Snehal. Sorry I didn't . . . they didn't make more progress on their project."

"So am I, but it is not your responsibility to make my son do his work, it is his. I am afraid I must get Snehal back home."

"I will walk you to your car," said Martin, because he wanted to ask—

"What did you make of the report?"

"Haven't seen it. SETI does not have the best equipment—"

"The SETI alert said the NRAO had also detected it."

"I wonder which telescope? Well, perhaps we will find out in the morning. Good night."

"Dad, can I come back tomorrow night to work on the project with Patrick?"

"To play your game I expect."

Martin knew Nathan wanted to go, but he blurted out, "My wife will be here—but Snehal is welcome."

"Thank you, but it is time for Patrick to come to our house, if he is interested."

"I am!" shouted Patrick from the door.

"Well, perhaps the wives can confer. Again, good night and thank you."

Was Nathan annoyed? He was unfailingly polite, but there was a hint of impatience. If only I knew how to read other people, Martin thought. I am the space alien.

The phone rang. Becky never answered; instead, she took advantage of the way phone messages were transcribed accurately and emailed to her.

"Becky, this is Ralph. Wanted to give you a heads-up. We have picked up a radio signal on multiple telescopes at multiple wavelengths, and it appears to have a complex pattern. The signal is not consistent with any astronomical pattern we have seen before. The SETI group is already spreading the word that ET might be calling. Unlikely, but at this point we have no alternate explanation. Hope your Monday is off to a slower start than mine."

She was on the phone immediately.

"Ralph, thanks for the heads-up. How widely have you disseminated this?"

"We have shared it with other astronomers who can check our observations and with you. But SETI has sent out an alert."

"At least it is a day late to make the Sunday talk shows. But OLPA will be after me for a response pronto."

"OLPA?" Good thing she remembered what this stood for. So many acronyms had come to be names that she knew which agency or committee they referred to but not what the letters stood for.

"NSF Office of Legislative and Public Affairs. Thank God I am not supposed to talk to the public—they do it for the NSF, but they will want to be told everything. Listen, I am going to use this as an excuse to make a trip to Charlottesville so we can confer in person." That would give her enough time to make sure that Patrick was getting on the bus, give Grandma a break, and make sure the world's only exopsychologist was on his meds before the press thought to call.

"Meanwhile, link me to everything you can share. I want to be up on all of this quickly as possible."

"Becky, the WOW signal appears again in the data."

"WOW?"

"A powerful but brief radio emission at the frequency of hydrogen that was picked up by a radio telescope called the Big Ear in 1977. Philip Morrison thought an alien intelligence might use this frequency to signal its existence."

"Oh yes, I remember now."

"The same brief burst appeared several times in the radio—emissions? Signals?— detected over the last 18 hours, from the same part of the sky, near the globular cluster M55."

An hour later, Alonzo, the Deputy Division Director, leaned on her door. She liked the way people just stepped in and out of each other's' offices at the National Science Foundation (NSF). Alonzo was competent, cheerful, and hard working. "Hi, Becky. Sarah is getting a few people together to talk about—well, about whether ET is calling." He laughed. "Or, more likely, whether we are looking at a novel phenomenon that warrants a special solicitation."

"You sent to escort me?"

He laughed again. "You know my job is to herd Program Officers. Come at 8:30."

"Not much prep time."

"No worries. We are just trying to get ahead of an emergent challenge—or opportunity."

Becky asked whether she could telework Tuesday to talk directly to the NRAO and make sure Patrick was OK.

"No problem." Alonzo had been great about accommodating her family situation—one of the reasons she loved the NSF.

Sarah had a conference table and a view out over the buildings in Ballston. Better than my monk's cell, Becky thought. Two Program Officers were already at the table.

"Jim Clark from the OSTP is on the line, so say your full names for his benefit."

The White House on the line . . .

"Becky Olson, National Radio Astronomy Observatory."

"Mohammed Seradeglin, Particle Physics and Cosmology."

"Josh Epstein, Secure and Trustworthy Cyberspace."

"Alonzo Charles, DD, Astronomical Sciences."

"Jim Clark, Office of Science Technology Policy, the White House." Refreshing to hear full titles of programs instead of acronyms—must be for Clark's benefit. "Thanks for coming together on such short notice. I need an update on the evidence of unusual—physical phenomena—from outer space, which some groups" he did not mention SETI by name "are already claiming might come from another civilization."

"Becky?" Sarah turned to her.

"Ralph Overman, NRAO director, called this AM to say he was getting unusual radio signals from an unknown source. I have looked quickly at the incoming data, will go down to Charlottesville tonight and talk to Ralph and others at the NRAO tomorrow. There is a pattern to the signals, but no one has been able to decipher it. The only recognizable part is a repetition of the old WOW signal in the sequence."

"WOW?" asked Jim.

Becky explained the original signal.

"Hmm, same place, same signal. The rest comes from the same area?"

"Yes."

"Anything there?"

"Not that we can detect."

"Is Hubble looking?"

"Don't know. That's the Space Telescope Science Institute, funded by NASA."

"One of my colleagues is checking with NASA. Be nice to get all of these observations coordinated." Clark sounded a little frustrated—probably young, not used to bureaucracy.

"If there is anything we can to help with coordination" said Sarah.

"Find out what you can and we can compare notes." Alonzo nodded at Sarah and left—obviously, to jump on this.

"Mohammed, why don't you brief Jim on neutrinos?"

"The Supernova Early Warning System has detected an unprecedented number of neutrinos coming from the same location near M55. This would be the first time SNEWS has detected an event of the magnitude that might be produced by a supernova, but, given our experience with the system, this supernova would be visible during the daytime. Furthermore, the neutrino emissions do not seem random; we are trying to find a model or relationship that fits them into a pattern. Problem is, we know we are missing most of them."

"Because they have almost no mass?"

"Right, we detect a vanishingly small fraction of the neutrinos that pass through our detector. Last time there was a nova in our galaxy, we detected nine."

OMG, thought Becky. There could be something really interesting out there. She began to get excited—this would make astronomy a high priority.

"Any help you need?"

"We just need time."

Josh spoke. "Cyberinfrastructure has funded PIs who have the intellectual and computational resource to deciphering codes. If there is a message in the neutrino flux, we might be able to help find it."

"Thanks" Mohammed replied. "Right now, we are still doublechecking all the measurements to verify all of the neutrino observations."

"I think I could help persuade a few PIs to work on the radio waves, too. Clearly, there is something unusual going on."

Imagine, all of these programs and PIs working together. Takes the threat of aliens to make us collaborate, Becky thought.

"OK, Sarah, thanks for putting together a helpful meeting so quickly. Keep in touch as you learn more. Goodbye." Clark was off the line quickly.

OK, everyone, thanks from me, too" said Sarah. "Becky, can you stay for a minute?"

"Sure."

"You are going to Charlottesville?"

"Yes, I get my son for the weekend."

"Is it too much to ask that you talk to your husband about business? He is the only person I know who has written and lectured about using signals from another civilization to infer what they might be like."

"Yes, he is good at science fiction" Becky practically spat the words out, and then regretted it.

"OK, well if you are not comfortable—"

"No, no, I have to talk to him anyway." And I have to confirm that he is on his meds—if not, I get Patrick, though Lord knows what I would do with him in Ballston. Maybe the aliens will send a solution.

Becky drove up a tunnel in the trees, punctured by an occasional house. Ballston was just a place to sleep; Charlottesville was home. In less than a month, she would have to decide whether to stay on at the NSF or go back to the NRAO. Then she would have Patrick for the week, and Martin would get weekends. That would be great, but she felt like she was just now really getting up to speed at the NSF. It gave her a central position in this business of—what? No progress on decoding any kind of message, but the signal had now been detected in X-rays and even in gamma rays. So it was either the weirdest astronomical source ever detected, composed of dark matter with a supernova in the middle and possibly a black hole, too. Or it was aliens unimaginably advanced. Or someone with an expensive sense of humor had figured out how to simulate all of these sources and pretend they came from the direction of M55.

She pulled up on the street because Martin's car and her mother's filled the driveway. Thank God Mom was there to help manage the transition; without Mom, she would have had to leave the NSF early.

There was another car, too—a shiny new black Volvo?

The old place needed a hug. There were still leaves in the gutters, peeling paint on the walls and a lawn as unkempt as Martin's hair. Somehow that was comforting—one corner of the cosmos that still looked normal.

She took a deep breath, and walked in. Patrick and Snehal were on the computer. Snehal was a straight A kid—a good influence on Patrick, whose performance was more uneven. Should I have sucked it up and stayed for him? No, just too much to deal with—

Martin was sitting at the old table right outside the kitchen, talking with Dr. Krishnamurti from a position where he could watch the kids. Damn Martin. She couldn't tell him to leave while a colleague was there.

"Oh, there you are, dear."

Mom gave her a big hug and Becky returned it with a bit more desperation than she had intended.

"You all right? Can I get you a special cup of tea?" A special meant just a soupçon of bourbon.

"Yes, a lot going on—"

"Hey, Patrick, it's your mother."

"Hi, Mom" said Patrick, not turning around.

Martin and Dr. Krishnamurti stood up. "Becky, if you need me to leave, Nathan and I can go somewhere else. You can guess what we are talking about."

OK, score one for Martin. It would be better to have her 'exospychology' chat with someone else present.

"Hope you don't mind if I join you?"

"We would be delighted," said Nathan, inclining slightly in her direction. As she sat, her mother came out with a cup of tea for her, and the pot to refill the others. Ah, that means I have the bourbon.

"Dr. Olson—"

"Becky, please."

Nathan smiled slightly. "Very well, Becky, you probably know more than we do."

She updated them on the X and gamma rays, saying everything was still subject to careful verification.

Martin leaned his head on his elbows with his hands over his eyes—an irritating behavior he said helped him think. Nathan leaned back in his chair, looked up at the ceiling as if gazing at the heavens.

"I can think of no astronomical object that could make those signals—it would have to be a combination. The signals are all from the same place?"

"That's the current best guess," Becky said.

"No one—no Terran—could fake such a signal," Martin said, opening his eyes. "If it is an alien civilization, they are unimaginably far beyond our state of knowledge." He spoke with authority—not the hesitant, erratic Martin she knew.

"Hey—hey, Dad, come look! A new civilization!"

"Pardon me for a moment—" Parenting was a constant series of interruptions.

In their small house, the living room and dining room were combined, so the computer table sat only a few paces from the dining room table. Meant she and Martin had occasionally been able to have a glass of wine while Patrick was engaged in some IQ-lowering activity.

"So Nathan—is this contact?"

Nathan sat Buddha-like, concentrating on the screen. Had he heard her?

"This civilization leader has no picture—the rest all do" Patrick said.

"But the response buttons are the same?" Martin asked.

Nathan rose and stood just behind Snehal.

"More tea, dear?" asked Mom, and then she leaned closer to whisper. "What are they doing?"

Becky touched Mom lightly on the arm and stood up to find out. She stood behind Patrick, with a little distance between her and Martin. Blank frame in the middle of the screen with some buttons.

"I'm going to try to make friends." Patrick pushed a button below the empty window that said "offer friendship." I thought all these games were about blowing up bad guys, Becky thought.

A message popped up. "Offer accepted."

"They never do that!" said Patrick. "You always have to give them something."

Something was happening in the empty window in the middle of the screen. Stars appeared—then they started to fly by. The window expanded to become the whole screen. The sensation of falling toward something—a fuzzy shape—

"Looks like a globular cluster," said Martin.

"Yes," said Nathan.

"Game has never done this before!" said Snehal.

"Is it new?" asked Martin.

"Yes."

"Maybe a special feature?"

There was a faint kind of—music? Or just static? Accompanying their fall toward the globular.

"You got this hooked up to the good speakers?" Martin asked Patrick.

"Yeah, I think—let me see."

Martin turned to Nathan. "I'm a Neanderthal compared to these kids."

Nathan smiled, but he never took his eyes off the screen. Suddenly the volume jumped. Snehal, still at the computer, turned it down.

The sound was less like static and more like some kind of modern digital music—Philip Glass and John Cage speeded up and combined. Martin's nose was almost touching the screen—

"Dad!" He pulled back abruptly, rocking slightly. Oh, God, I hope he is on his meds!

"It is a message. I have been almost hearing it—"

He is hallucinating. She put her lips right against his ears. "Have you been taking your Abilify?" He touched her arm.

"Bless you for still caring about how I am doing," he whispered. His eyes did not move.

"The glob is M55. We are going to pass by it. We are going to the space where the WOW supposedly came from."

If I were religious, I would pray, thought Becky. She heard Mom bustling in the background.

A dark spot grew—and there was a square of light in the middle. They fell right through it and into the light of a sun whose edges seemed to be constantly moving, shifting colors. They stopped. The strange music seemed to surround them.

"If only I could decipher—oh!"

Mom was right behind Becky again, slipped something in her hand—a plastic container. Abilify! Have to get Martin aside.

"I can move," Patrick shouted, and sure enough, their field of view rotated away from the bright sun, thank God. It was the inside of the sphere that was full of colors and shapes that were strange, unsettling—but impossible not to look at. Do I need Abilify, too? Becky wondered.

A silver disc grew in front of them. Wait, that can't be—

"Camera is on," said Snehal.

The screen flickered and the picture frame was back—with all of them in it, staring at the screen.

"Contact!" said Martin, standing up. "Or an incredibly sophisticated modification to the game, with inside knowledge—perhaps SETI folks are involved. Two hypotheses—the game has been infiltrated by alien intelligence or the game has been modified. Any other hypotheses? Oh" looking at Becky. "That I am hallucinating. You all know that I am on Abilify to prevent delusional thought."

"I didn't," said Snehal.

"I should have told you, sorry—I guess I was a bit embarrassed because I like you."

This sounded a bit manic?

"OK, so let me review what I saw and heard." Martin gave a brief account.

"Did you all see the same things?"

"Yes," said Nathan firmly. The boys chimed in. Becky hesitated.

"Ah—Becky—in your hand. Is that the Abilify?"

"Yes, ah—"

"Give it to me, please. I have not taken my dose for today. Let's make sure I am on it. What did you see and hear?" Becky shook her head as if to clear it. "Same as everyone else."

"Collective hallucination? Doubtful. I will take my Abilify now, but let's eliminate the psychosis hypothesis."

"Patrick, you and Snehal continue with the game to document what else happens with this civilization. You each take a turn, then the AI—and the—ah, new Civilization—take turns? Can you see what they do?"

"No, we can only see the results."

Martin stopped, frowned.

"Continue," said Nathan. "Show us the results of each move by this—interesting civilization. Do we have another computer?"

"I brought my laptop" said Becky.

"I have one, too," said Martin.

"Excellent," said Nathan, taking over. "Becky can you keep up with the incoming data on the signals from your laptop? Has anyone done an audio recording of the incoming radio waves? Let's try to match it to what we heard just now."

Martin, you look up information on this game—on whether it has been modified recently, perhaps by a sophisticated programmer who has access to the radio data. I will go to my office where I can use much more powerful computational resources, in case we need them. Martin, look for me online."

"Good plan. Mother—" Becky was not sure she liked the way Martin had adopted her mom, and she had adopted him. "I am afraid this calls for coffee, and probably some calories, too, if I can trouble you—"

"I'm on it."

This is too much. "Martin, you know I have custody of Patrick over the weekend. You have to leave."

Everyone stopped for a heartbeat or two.

"Mom, I like it when you and Dad are together!"

This is exactly what I wanted to avoid. Now I am the bad person.

"Nathan, I have an office down the hall—how about we both go in?"

Snehal and Patrick looked at one another. "We want to come, too."

"I wonder if we are the only species in the galaxy that does this, or whether it is a stage that all go through? We may have taken one step toward solving this mystery tonight." Martin was a master at changing the subject to avoid confrontation.

"Patrick, you know the rules," and suddenly she was crying, which made her angrier. "Damn you, Martin, you always back me into a corner."

"I think this time you have backed yourself into a corner, dear." Even her own mother took his side. They probably hope I will storm out. Well I won't. I know my rights.

"Patrick, you and Snehal have a job to do, right here. Stay on the game. When we get to the office, I will ask you both for a report. Then we can try a new multiplayer online game and see if this unknown user joins with this strange civilization. Is that OK with you, Nathan?"

"I will pick you up in a couple of hours, Snehal."

"Can I spend the night with Patrick."

Nathan looked at Becky. I am back in charge. "Sure, happy to have him."

For the first time, Martin felt detached from the argument with Becky, as if it was a pattern that she was locked into but he was not—not any more. He still cared very much about Patrick, and even felt sympathy for Becky. But he had glimpsed something that might show their whole way of being was just a way of being—that there were other ways of being that might be totally, well, alien—

"You are deep in thought, my friend. Are you all right?" Nathan held the car door for him.

"Nathan, you are a perfect gentleman. I have always admired that. I would apologize for the scene in there, but that is what it means to be human."

"You are not angry, or hurt?"

"No—I would have been, but that music, that sense of something wholly beyond this little world we all construct as we go through our lives. Perhaps it is what the truly, deeply religious feel."

"I have always felt religious, in that way." Nathan's eyes were fixed on the road as if he were looking beyond it. "We are free in ways that most of us do not imagine."

"You—you are wise. I never knew—"

Nathan laughed. "The corridors of a department are not exactly temples of wisdom. It is where we do our ordinary business—punctuated by an occasional idea, or a new piece of data—moments when we see what no one has seen before. Then we return to worrying about how this discovery will lead to prizes, awards, titles—even sometimes pay."

"Well as one of the underpaid peasants down the hall, I would at least like a grand title to put on my door." There was a moment of silence, and then they both started laughing so hysterically that Nathan had to pull over.

Her cell phone beeped. *Dammit, I meant to turn it off.* There was the faintest hint of light through the window. *It better not be Martin.* As she reached for the phone, she saw his Abilify on the nightstand. *Damn!*

"Becky?"

"Ralph? At this hour?"

"Becky, I can't talk over the phone. Can you meet me at my office? I will get us both coffee."

"Uh, sure." *Something big. And secret. I'll leave a message and tell Martin to pick up his meds? No, then he will be back in the house. Quick shower. Throw on something clean, professional—damn, rumpled, have to iron it while the coffee brews.*

Mom came into the kitchen while Becky was praying over the coffee maker, blouse over her shoulder.

"Do you want your son to see you in your underwear?"

"Would you care if he saw Martin in his?" *Deep breath—let's not punish Mom for supporting Martin.* "Sorry—just got a call from the NRAO; they need me in for a meeting. Now."

"You poor dear. Won't they even give you a Saturday? Here, I will iron the blouse if you go up and put a bathrobe on."

"Oh, and Martin left his meds. I better get them to him right away." *Or else he might freak and I would get custody of Patrick?*

"I'll take care of it, dear." *The unholy alliance—my mother and my psychotic spouse. All right, if she wants to worry about it—*

"Thank you."

Ralph was on his cell phone, pacing when Becky came into his office. He pointed to a Keurig. "So you want us to stop receiving the signals?" *I am going to need strong coffee to process this.* She picked out one of the little containers that said Bold on it, rammed it into the Keurig, found a Styrofoam excuse for a cup, and hit the button.

"We can only stop receiving if we shut down. How about if we isolate the IT that receives them—cut it off from all external systems?"

When she took the cup out, she grabbed it too tightly and it spilled. *Crap!*

"OK, look, our NSF PD is here, I will discuss it with here. OK, OK, I hear you, just give me a minute, OK?"

Ralph put his cell down. Becky took a careful sip from her cup without taking her eyes off Ralph.

"OK, Becky, I know you are with the government but I am supposed to remind you to keep this top secret. Looks like these radio signals contain a virus that is infecting our computer systems."

"Maybe that explains what happened to our own computer last night." Becky described falling toward a point near M55, the strange music that could have been the radio signal, and the final picture of all of them.

Ralph sat down. "Could be a prank, though it is an interesting coincidence. But cyberattack recognition software is sending alarms about a new threat and the traces lead back to our computers and SETI's. We can see no viruses on our systems, we show no attacks on us, no attempts to penetrate—just the detection of this signal from space. So Cybercommand wants to shut us down."

"But, but—other nations, other astronomers are receiving it, too. The barn door is open."

"Yeah, they still want us to close it—tight. What do you think?"

"I'd like to be funding research on this signal, where it came from, how we can decipher the pattern, etc. If we could isolate a part of the NRAO's system—lock it off from the rest of the Internet—"

Nathan had let Martin inside the sphere with the sun in the center. The wall was a giant mirror, focusing the light on Martin, blinding him, but he had to watch his hand trace the complex patterns of the music across the light. He almost understood it.

"Dad, Dad, you still there?" Patrick's face appeared on the sphere and Martin dove toward it.

He had fallen asleep in front of the computer. "Hey, yeah, still here, dozed off. Time you were off to bed."

"The alien player is back on. There is a message from—him? Her? It?"

Martin sat up straight. "What's the message?"

"Restart the game."

"Do it!"

"Whadja find out about mods to the game?"

"Didn't see anything that resembled our unknown Civilization on the official site, nor among the user's groups. Posted what we saw in a new thread. Will check later to see if anyone picks it up. Snehal still there?"

"He's sleeping on the floor."

"Where's your mother?"

"She just left for some kind of early morning meeting. Wait, Grandma says you left your Abilify."

"Sh—oot, yeah, gotta get it before Mom comes back, but I see my invite. Let's look for the Civilization again. And Patrick?"

"Yeah?"

"You have to check me on everything I think I see and hear. Snehal, too, when he gets up."

"Got it. Is Snehal's dad there?"

"No, I'm the only one in his lab; he gave up on the new Civilization a while ago and went off to bed." Martin was using one of the high-end Macs in the lab; Nathan had given him the password and said it was his station now—he could use it anytime. Much better than a frackin' title. The half dozen other workstations and computers looked like small stele as they disappeared into shadows, with a supercomputer in the corner blinking and breathing. Martin had turned the lights off and worked from a reading lamp he had found. A tomb of the ancient astrophysical modelers, hidden deep in a faceless building.

One computer was on, showed M55 off to one side, a great ball of diamonds riding on a pool of darkness. Was there a spinning sphere containing a star in that darkness?

"Patrick, you and I may be making the greatest discovery in human history—or I will be laughed at forever. What I don't get is how we could converse with aliens thousands of light years away? Not possible."

"Game's up now—you have an invite?" Patrick knew his job—*keep Dad from flying too far from reality.*

"Yes, but it's not from you. It has the picture of the four of us on it."

"I'm getting an invite, too. Wake up, Snehal—you have an invite waiting for you!"

Once the game came up, it showed three human players—Martin, Patrick, and Snehal—and one AI.

"I have figured this out, lads. I will write it in the chat window so you and the so-called AI can see it."

"Huh?"

"Give me a second."

"Congratulations on a brilliant hoax. You must be a sophisticated programmer who has access to the latest astronomical information. You created a very clever AI that plays a recording of the unexplained radio signals from near M55. I don't know what you had planned for this game, but I need sleep, so tell me or I am dropping out."

Who would do this? Only one person he could think of—Nathan. It would be a wild prank and he now knew Nathan understood the stupidity of taking oneself too seriously—*which I am guilty of.*

His cell phone rang. He answered it automatically, watching the screen.

"Martin." Becky was probably in on it, calling up to laugh. "Are you and the boys and Nathan still playing that game?"

"I'm about to end it—the alien civilization is a very clever hoax, but I've seen through it."

"Oh, good, but the radio sounds from near M55 came out of our computer last night, I think."

"Yeah, Nathan and I checked it. Whoever played this trick recorded the incoming sounds."

"Is Nathan there?"

"No, he left a few hours ago."

"Damn, we need his help. How can I reach him?"

Martin read Becky the number off his cell phone. Although the prank was funny, it also made Martin a bit depressed. He thought he had found a soul mate in Nathan, not another person who would play tricks to see if Martin could be fooled.

"Was that Mom?"

"Yeah, trying to reach Snehal's dad, so I gave her the number."

"You really think this is a hoax, Dad?"

"Yeah."

"Well, let's play anyway. Snehal and I took our turns. We are on Earth."

"Both of you?"

"Yeah."

Martin saw it was his turn, and he was also on Earth. The first move was to build a spaceport, which he did automatically, without thinking. The strange radio transmission noise came on, signaling it was the hoakster's turn. The screen popped up with their faces in it, and under it a game message that asked if he wanted to share maps. *OK, I will play along, look for a chance to expose whoever it was—apparently she or he was not going to answer his chat.*

His map shifted from the solar system to a globular, probably

M55. There was a blinding flash—a supernova, sending light and heavy elements across the globular. He zoomed to a star on the edge with a solar system. The fourth and fifth planets lit up, and little lines zipped between them when they came close together in their orbits.

"Two inhabited planets around the same star—I think the lines are space travel," said Martin.

The yellow star started to fluctuate slightly.

The globular was moving closer to the center of the Milky Way. The blue, brown and red on the fifth planet's surface changed to gray. Lines came from the fourth planet, along with their larger moon, which was rapidly reduced to rubble and added to the surface of the fifth planet.

"Maybe two civilizations cooperating? They are terraforming the fifth on a huge scale, I think," said Martin. "Perhaps they are doing this in anticipation of—"

The solar system was pulled out of the globular, capturing another star. The fifth planet used the disruption to exit this binary system, ending in a space without other nearby stars. Little pinpoints of light shone out from inside the planet.

"How could that be done? Fusion?" Martin snorted. "This is a beautiful fantasy."

A gray circle spread from the planet, and then circles of different colors emanated.

The screen suddenly shifted to Earth in its solar system; the gray circle passed through it, and then the many-colored circles. Then lines started to spread out from Earth. "The circles are signals. We know about the ones coming in; how about the ones going out?"

"Hey, Martin, you have been here all night?" Nathan was wearing dark pants and a white shirt, but he had shed the tie and jacket.

"Yes, and I need coffee or a bed. Did Becky reach you?"

"Yes. Are you playing with Snehal and Patrick?"

"Hey, Dad."

"Snehal, how are you." Nathan leaned close to the computer. "I think you have a soccer game this afternoon."

"Dad, we saw where the aliens came from!"

"Interesting—"

"I think it is a hoax by a clever programmer who knows a bit of astrophysics and has access to the radio emissions" said Martin.

"Good hypothesis. Did you check the game discussion boards?"

"During the night. No mention."

"Let me make some tea and talk, then you must get some sleep, and Snehal and Patrick, you must, too."

"Hey, Dad, I just allied with the aliens and they gave me fusion power!" Patrick said.

"Great. Tell you what, you guys continue with the game for a bit more, take good notes—I am going to follow Nathan."

Martin opened the door just in time to see Nathan go into his office. He ran down the hall for the joy of moving.

"Nathan, sorry to barge in but I thought we might talk. Can you tell me what you learned from Becky?"

"No, but I will as long as you promise you did not hear it from me."

"Hear what?"

"Good. The cybersecurity experts think that the radio signals, when decoded, become a virus that is infecting all of our systems that it could reach."

"Including your computer down the hall?"

"One proposition is to back up our systems and wipe them."

"We can't do that with all of them!"

"Exactly what I told them—ridiculous idea. And I refused to do it with mine."

"Too late to close the gate." Martin was drifting in and out sleep. Something he had to tell—

"Nathan! I know I said the game module was a hoax, but it showed signals coming to Earth—then ones going out. No idea of the time interval because time seemed to shift from billions of years to days—very bizarre."

"My friend, are you sure you wouldn't prefer sleep to tea?"

"Yes, but I have to get my Abilify and make it to my apartment."

The tea was blissfully strong. Plus Nathan had called him "friend."

"A message is going out now."

"A message from what?" Nathan asked.

"From whatever is in our information and communications systems."

"How do you know?" *OK, here is where I maybe lose a friend.*

"I just know—or I need Abilify—or both."

Martin was sitting next to one of the aliens—so close they were touching. Two outcasts, one on his home planet, the other light years away. He looked toward his new friend and saw a nebulous glow gradually embracing him. Someone was pounding on his door.

"Uh—hey, just a minute." *Must be the guys from the funny farm "coming to take me away, take me today"* With apologies to the Beatles

"Nathan."

"Martin, can I come in?"

"Um—look, this place could be an EPA Superfund site—"

Nathan pulled him in, shut the door, and appeared to ignore the bachelor chaos around him.

"SETI bulletin. Communications satellites are beaming messages into space. No other confirmation at this point."

"ET phone home?" Martin said, hoping Nathan would laugh.

"Martin, I think you were right! But I don't understand why. If this is an alien virus from the region of M55, it will take about 20,000 years for the signal to make it back to its origins."

"The signal must be intended for something closer. Perhaps a fleet of ships at near light speed is following the signals and needs to be directed toward signs of intelligent life—though sometimes I worry about the amount of intelligence on this planet."

"A virus followed by an invasion fleet?"

"The problem is, we have to think like a different sort of mind. Our best clue is what the AI is doing. Hey, thanks for coming over to give me the heads-up. I gotta take a shower, get coffee, and think, think, think. This may be the greatest thing that has ever happened to our species."

"Or the end of it," said Nathan.

"Let us help you, Mom."

Becky looked up from the laptop. Patrick stood on the other side of the screen. *He is tall—amazing how it slips up on you.*

"You are here to spend time with me, right?"

"Uh, yes—but I have this—"

"I know. Work. I get it—this is important. Snehal and I would like to help." Snehal looked over Patrick's shoulder and nodded. "We were in the Civilization game—we saw strange things. Nathan and Dad are working on this. We are part of it, too. This will affect our generation even more—"

"Spare me the rest of the speech, especially the bit about my old, decrepit generation ruining the planet for yours. You're in."

"Let us start with the game—"

"Oh, I get it. This is a way to avoid homework, chores—"

"No, Snehal and I will clean up the kitchen first. The game was a means of communication—"

"Yes, with your father. This is my time!" She was surprised at the edge in her own voice. *Tone it down.*

Patrick came around the table and hugged her. For a moment she froze—when was the last time he had hugged her—then she wrapped her arms around him, and put her head against his chest and felt tears.

"Mom, I love you. I love Dad, too. Why can't the two of you get along? Why can't most of us on this planet get along?"

"I don't know—"

Patrick stumbled back, and put his hands up to his eyes. Becky jumped up put a hand on him. He pulled away.

"No—wait—I am here, and I am in Dad's apartment, too, where he was looking at his computer until something joined us—something that is looking for the answer to that question."

Oh, God—Patrick had his father's genes. She knew several teenagers who had become bipolar. Her computer started beeping at her. Ralph would have to wait.

Martin was at his laptop, but could also see through Patrick's eyes—see Becky, see the old house, feel the old question he had apparently bequeathed to Patrick—*why don't we get along? If we each saw through the other's eyes—*

The moment ended. He had to find out. He called Patrick on his cell.

"Dad, I was there with you—"

"And I with you, for an instant. Patrick, I have a hypothesis."

"You always do!"

"This has something to do with the aliens. I dreamed about and have felt a presence—utterly different, beyond my understanding, yet sympathetic, even loving. Must be what people feel when they know God?"

"I think I felt that, too. Just a minute, Mom! This is weird in a kind of wonderful way. Got to figure it out. Snehal is nodding. OK, Mom—here he is."

"Martin, what on earth—"

"Not from Earth, Becky—from thousands of light years away, or maybe it has always been here—"

"This is crazy, it makes no sense."

"Becky, before Einstein we knew that space and time were separate—"

"Spare me your goddamn paradigm speech! Why has this happened to us?"

"It might be happening to others—perhaps thousands, even millions. I am coming over. It is apparent we are meant to work on this together."

Got to tell this to Nathan—and suddenly he was seeing the lab. All the students had been shifted to monitoring events. Nathan was at a whiteboard, writing down each piece of information, trying to draw connections.

"Ah, here is a bit of data that transforms everything. I feel like Martin is looking out through my eyes. He cannot read my mind, I cannot read his, but we are looking together. This is not a human capability—somehow the signals and the cyberpresence are involved."

The students stopped working and their eyes widened.

He accepts it instantly, thought Martin, and immediately felt Nathan's response. "Of course. One accepts the reality of the experience, then tries to understand it." Martin was not sure whether he heard words, or translated a thought into words. If he did not fight it, it flowed.

"I am going to work with my son and your son."

"We will stay connected, through this means and more conventional ones." He felt a bright splash of laughter. Then Martin was 100% back in his room, shaking slightly. His whole system, his whole way of thing, cried out that this was impossible, that he must have imagined it. He had to call—

"Nathan! Was I just there?"

"Yes—and you left too quickly for an update. *The Washington Post* just broke parts of the story. Let me read you highlights."

Thank heaven he wants to read it—I am not ready to look through his eyes again.

"Is ET calling?

Signals from outer space may be from another civilization"

"Ralph Albright is quoted: "There is definitely a strong and complex radio signal from a region near M55 where there is no visible source. We have been unable to decipher the signal so far.

"When asked who is working on it, he fingered the Department of Homeland Security.

"A DHS spokesperson said they were investigating the signal just to make sure it represented no threat to national security. 'We can't be sure this is from outer space. Perhaps this is a clever hoax designed to probe for weaknesses in our cyberinfrastructure.'

"The Chinese were more blunt. They were sure the US was conducting some kind of probe of their systems. The Russians are growling at everyone. No one referred to an alien virus or to signals being sent back into space.

"Apparently, SETI does not have enough cred to make the *Post*—yet."

The doorbell rang. It's him, thought Becky, and tried to be ready. Patrick ran to open it.

"Dad!"

"Becky, sorry for the intrusion. You see today's *Post?* Sounds like Ralph has had to give the ball to DHS, and I suspect they are saying a lot less than they know. You know anything, Becky?"

"Not that I can tell you," she said, suddenly angry. "You should not be here."

"Dad, maybe you and Snehal and I could go to Nathan's lab?" Patrick had moved closer to his father.

"No—this is my weekend. We agreed, Martin!"

"Becky, this is a once-in-a-civilization occurrence—contact with another sentience. Can we suspend the rules for an afternoon? It's not like this is ever going to happen again."

"No!"

"Becky—"

"No, Mother, no! This is why we have a court order—to prevent my having to be a bad cop."

Why? Martin perceived the thought, and then translated it. *We want to understand. Maybe we did this, long ago.* Again, the thought was wordless, but clearly not his—and he had to translate it.

"Earth to Martin. Did you take your meds?" said Becky.

I—I need to sort this out, he thought back, hoping whatever was in his mind would understand.

I will watch.

"Right, sorry, there is a lot going on. Patrick, I will go Nathan's lab. We can connect from there. Be back online in a half-hour. Oh—meds, yeah, I took them. Mom, Snehal—" he looked at his wife who was staring daggers at him.

"Exit, stage right," and he was through the door, into the car, where he took a deep breath.

What do you mean, why?

Why do you interact this way?

You mean Becky and I?

I can see the relationship as you see it, and it does not make sense—you are both drawn to her and repelled by her like a planet whose orbit takes it far enough from its sun so that it might escape.

Martin saw the planet going around the star. The communication is instantaneous—it just takes me time to turn it into words and pictures.

But you are not a planet and she is not a star.

Then Becky's face was at the window.

"If you don't move, I will call the police!"

He was surprised that he did not feel the usual spark of rage. She was doing the best she could in what she thought of as a difficult situation, without seeing that her own actions made it worse. He rolled down the window.

"I'm sorry. I'm a bit overwhelmed by all of—aliens? Really? Wish we could all work together. Goodbye, my love." He was not sure where the *my love* came from. It just bubbled up.

She was sitting on the ground, crying. He moved to the passenger seat to get out so he would not hit her. He sat next to her on the little patch of grass between sidewalk and road, careful not to touch her. All around them were trees, driveways, and modest houses with mowed lawns and shrubs.

"I'm sorry I can't be what you want me to be."

She looked at him, red-eyed. "It's not your fault that you're insane, but it is your fault you don't recognize it, and I can't deal—I need a partner who is an adult, not a liability."

This opening is what let us communicate. You look while you are acting. You do not take these selves, these made up worlds, for granted.

"Look, Becky, can we at least work together on this alien—whatever? On something that transcends us?"

Nothing transcends the need to understand one another.

"Yes," said Becky, "but from a distance."

What is this distance? You are all squeezed on one world.

Martin stood up suddenly. Patrick was standing behind them, looking worried.

"Hey—" Martin put a hand on him, and suddenly he saw Becky from Patrick's eyes—not just the eyes, the whole different way Patrick looked at her.

You still love Mom, but you are angry because she wants you to be something you cannot.

I think I have an alien inside my head, I am inside your head, I am both sane and insane. I am here, and I am also somewhere else—

Becky tried to stand. He helped her.

"I was wrong," he said. "Our relationship—the three of us—every relationship—is the cosmos."

Becky sighed. "Meds and therapy are what you need."

"A way to avoid seeing the real problem—the real opportunity—"

Both of their cell phones went off at the same time.

Martin was glad Nathan offered to drive: the alternative was a lunatic with aliens in his brain and a car that belonged in a junkyard. Nathan agreed to wait until Martin's 8 o'clock section was over.

"Good morning." There were supposed to be 30 students, but less than 20 were present. 8:00 a.m. classes were ordinarily fun—you could say just about anything and no one would be awake enough to notice. But this class would be recorded for his other sections.

"Tell me what you know about these purported signals from outer space."

Silence. Wait for it—

"Science News Online says that the radio signals contain a complex pattern that cannot be explained by any known physical source. There are—patterned signals in other wavelengths from the same source, much simpler. When Hubble was pointed at the source, nothing there."

Maria, one of the few students who spoke up and whom he therefore recognized. A dark-haired beauty, poised and thoughtful.

"Thanks, Maria. Note she is consulting her laptop as she speaks, a practice I encourage, and that she found a credible source. Anyone else have additional info?"

"Russia thinks we are making the signals to invade their cybernetworks."

Another student who spoke up often enough so Martin knew him. Hipster, tatoos, face jewelry, slim-cut suit. "Job, Justin. Anyone else?"

"Could be the Russians or Chinese are getting into our cybernetworks." Crew-cut dude in the back. Martin couldn't remember his name so he said, "OK. But how did they arrange to make the signal come from a point near M55?"

"Maybe they didn't—maybe they took advantage of the signal to attack our network."

"Good point. Where'd you hear that?"

"Fox News." *Of course.*

"So, you all have drafts of Anticipatory Governance scenarios. How many of you included contact with another civilization in your papers?" Martin thought there were two, but couldn't remember which sections—

One student raised her hand. "Marte, do you mind sharing yours?"

"Maite, Professor." *Brown hair, broad shoulders, intense—maybe a swimmer?*

"Sorry."

"I thought alien civilizations were already observing us, collecting data, and would contact us when they thought we were advanced enough—if we ever got there. We might blow ourselves up first, or wreck the environment so thoroughly that we had to drop back to simpler ways of living—after a population crash."

"Yes, that was a brilliant paper. Does what is happening now change your scenario?"

"Well, if this really is contact, no. Maybe it is time."

"Great idea, Maite—write about it." *Improvising.* "Here is the question I want you to answer. What advice would you give our President and his cabinet about what these signals represent, how to find out more, and what to do about them? Imagine the response your question will get. That goes for all of the sections watching this video. I am off to DC to give my own advice to the President's Office of Science and Technology Policy. If you put your advice on the discussion board in the next couple of hours, I can use it to sharpen my own thinking. Gotta run."

Nathan's hands-off phone beeped.

"Jim Clark, again. You on the road?"

"Roger," said Nathan. "Martin is here."

"OK, to recap, the President needs to say something to the American people about this situation—soon. The Russians and the Chinese are pointing fingers at us. We are saying that our computers are hit, too, but of course we cannot supply details. You two are among the dozens we are consulting. So, give me your first take on what the President ought to say. You'll have another chance to do it in more detail when you make it here."

The thought popped into Martin's head almost instantly.

"I think the signals are sent by scientists, not soldiers. These—whatever they are—want to find out about us."

"How do you know?" *Because I have an alien in my head telling me.*

I am not telling, you are discovering.

"A strong intuition, at this point. I need to make it into a testable hypothesis—"

"Our systems are compromised and the president needs to say something now to prevent panic and possible war. There is no time for testing hypotheses."

Can you get to see this President—so I can see/feel/understand?

"See the President? Don't be ridiculous."

Nathan tapped the dashboard near the phone. *Crap—I spoke aloud!*

"What was that? You're fading."

"Let us talk amongst ourselves while driving. Call you back." Nathan broke the connection. "What was that bit about talking to the president?"

"It was a request from—the Other."

Not Other—I/we exist in relation to you.

"Does the—Other—understand that even if we could get to the President, convincing him is not the same as—that the President has limited authority?"

No convincing—explain—observe. There is what you call fear everywhere. Your President has destructive power. So do others. Why?

"I got that!" said Nathan. "Wish I had an answer."

"Really? I could be dismissed alone, but if two of us get the message and the message is the same . . ."

There are others on this planet—far away by what you call "geography." It is hard for you to make the direct connection even when your nervous systems almost touch. It is as yet impossible over distance. I can connect you all digitally.

Nathan called Jim. "Martin and I are in communication with an intelligence that is inhabiting our global IT system."

I am your global IT system, now.

As soon as Becky got off the train—which blessedly had terrible Internet and no cell connection—she saw she had messages from Jim Clark and Sarah, telling her to call. She tried first, but had to leave a message. Then Sarah, who picked up right away.

"Becky, Jim called and asked if your husband and Nathan have lost their minds. They talked about being in direct contact with an alien intelligence."

"I think they are both crazy—but I'm not sure—Nathan is more level-headed than Martin."

"Where are you?"

"At the Ballston Metro, will be at the NSF in—"

"Go to the OSTP instead. Jim is trying to get info for the President to make a statement. Keep me posted."

Her husband had screwed up her family life. Now he was wrecking her job.

She was rushed through the usual security check and into a room. Everyone was arranged in a semicircle around a screen that showed the President's National Security Advisor.

"The Other, the alien intelligence, has made a testable prediction." Martin was saying. Becky both wanted to hear what he had to say and to strangle him before he could open his mouth. "It can stop all the cyberattacks going on at this moment, though it will let them resume after an interval. It says it cannot solve our problems for us—it is here to observe, offer advice if we want it, and connect us gradually to the rest of the galaxy."

The National Security Advisor shrunk to the upper left-hand corner. The screen showed the ongoing cyberattacks, most arcing from China toward DC, but hundreds of others from and to points all over the world—including the United States to China and back. Suddenly, they all stopped.

"How the hell—" said Jim. "This is a secure room. We shouldn't even be able to see that here. I should have the two of you arrested for this gimmick. If it is not a gimmick, I should arrest you for stopping our cyberattacks."

"Let me check before we get too excited," said Mohammed. His phone rang. He cupped it, listened for a moment. Then he looked up. "Someone at Cybercommand who asked not to be identified said they got an urgent message to contact me just after the cyberattacks stopped."

"Now the Other will let them resume." The screen lit up with multicolored tracers. "The Other has asked me why we do this, and is searching my often scrambled memory for clues—with my permission. It is doing the same with other minds it has made connection with around the world, none of whom wish to be revealed at this point as it might put their lives in danger. I have agreed to make my name known—"

"What about Patrick!" Becky shouted, and realized she was also thinking about herself.

"Sir, the Hot Line is ringing."

"Ah, perhaps the Tsar has the same problem I do."

"Please do not call the President a—"

"Not to his face. But he is. Snaps his fingers and things happen, whereas I might as well sing into the toilet for all the good it does—"

The President went with his National Security Advisor into the secure room with the phone. "Idiotic anachronism," the President muttered as he took the receiver from a staff member.

"Yuri, what can I do for you?"

"How did you stop all the cyber probes from all countries? I know you won't tell me, but we find it alarming."

"You mean attacks. We didn't. You know we would have stopped everyone else but us."

"No—you stopped everyone to test your new system for stopping us while you continue."

"Have you talked to the Chinese?"

"They deny that anything happened."

"They would. We are trying to figure out what is going on just as you are. Have your top National Security person stay in touch with ours—"

"You will shut down our missiles next!" Yuri cut off abruptly. Probably slammed the phone.

"Katherine," the President said to his National Security Advisor, "you need to coordinate a response to the Russians to keep them from starting World War III. Better check with the Chinese, too. Get State and Defense involved, but you are in charge."

"Sir."

"Larry," he said to his Chief of Staff, "track down our alien whisperers and find out if they should be trusted or locked up—or both."

Martin tried to meditate like Nathan, but it didn't work. Becky was right—family came first. Why did he have to be the one who told the planet? Why wouldn't the others—

Because you have hate/fear of the unknown, and of each other. But you love also. Why not love all your little ones?

Wish I knew.

What will convince your species?

If you could reach them all—no, even then some would call you God, others a devil, others a trick. They must come to this in their own way. We must gently persuade.

"The Chief of Staff will see you now."

An FBI agent had escorted them to the White House, where a Secret Service agent took over. Larry Smith met them in a conference room. He looked to be 40 going on 60, prematurely gray—probably 20 of those years added in the last 3. "Office is too damn cluttered. Coffee, gents?"

"Always," said Martin. Nathan was smiling like a Buddha.

"So, the President is still planning to talk to the people. No press conference—a little online 'fireside chat' to reassure. If you are talking to an alien or aliens, then ask it/them what the President should say."

"That we should love all children as if they were our own," said Martin.

"Don't waste my time, gentlemen," Smith growled.

"If we loved all children as our own, we would not need cybersecurity," said Nathan, smiling.

"OK, that's enough—" Smith said, starting to stand. The door to the conference room opened. One of the many young people who gofer in DC said, "You need to come look at your computer screen, Sir."

"OK, you two wait here and drink that coffee—maybe you will make sense."

"I think this is going well," Martin said to Nathan who doubled over, trying to suppress a laugh. Martin took a hit of caffeine. He missed Patrick so deeply he could barely breathe.

I am sorry, son.

For what?

Patrick was with him. Martin felt a glow

I am trying to convince people who do not want to listen, instead of being with you.

We are together now.

I told the Chief of Staff—

I know. You are right that we should love everyone as ourselves. But there are so many idiots who hate—

It is so hard not to hate those who hate—that perpetuates the cycle of hate.

We used to hate, too—when there were three species, same planet—we know the history, the feeling is lost.

Martha was back. "You had better come in—the whatever-it-is has taken over the Chief's computer, and he is pissed!"

Stay with me if you can, Patrick. Where is Snehal?

Connected with his dad.

Nathan smiled at Martin.

The Four Musketeers.

In the Chief's office, the old Civilization screen had appeared and was zooming into M55—except it was not Civilization. Martin realized he was supposed to narrate.

"OK, this is the—aliens—showing where they came from. One intelligence on a planet on the edge of a globular cluster—"

"What's that?"

"Essentially, a big ball of really old stars—so this alien race or species is very old, too." A creature like a brightly colored octopus appeared on the screen; its skin looked harder, like leather, and its tentacles ended in different appendages clearly adapted for different functions. Then a dark net started to grow over the octopus's skin, and it started to spin and tumble wildly, its tentacles waving in all directions, until it went limp.

"This black net turned out to be another intelligence that grew up on the land and migrated to the water where the octopus species lived—"

"Look, I got a dozen fires burning. Can't someone get this fairy story off my computer while you tell me the punch line?" Smith deliberately turned away from the computer. The gofer was fumbling with it. Martin looked over Smith's shoulder at the screen, which was defying all efforts to restart or turn it off.

"Get Christophe! If he can't fix it—OK, the two of you might as well brief me, but you need to explain to Christophe how you did this."

"We didn't. I am narrating what is being told to me by a being that does not use language but makes it possible for me to understand the story as it is unfolding. Nathan, you can correct me if I get it wrong?"

"I'm with you."

"Did the two of you make all this up?"

"Judge for yourself—but we did not." The visual story had stopped, and only resumed when Martin talked.

"The octopus and the microbial net eventually combined into a symbiotic organism. A third species hit the planet, went to war with the symbiote—except for a small number, who worked with the octopi to combine genetic material systematically, creating a new species that was a combination of the three. This species eventually dominated the planet and created the technology to move away from the globular in a sphere with a star inside. Now the three-into-one species went from having one body to being able to create any form it/they wanted—which is why Nathan and I saw crazy shapes when we went inside the sphere—"

"Wait, what—ah, Christophe." A tall, pale man with a ring of gray hair around his dome took the gofer's place next to Smith's computer.

"Nathan and I were playing a Civilization game with our boys and suddenly we were swept to this planet and went inside the shell."

"I can't believe I'm listening to this. Cut to the punch line!"

The screen grew brighter behind Smith and he turned back to it, cursing under his breath. "The new species made of three began to explore by sending out satellites, nanoparticles and, yes, a complex program on multiple wavelengths." The screen showed lines marking the routes. Some went into the cluster, others went toward the center of the galaxy, and still others went out toward the wilderness where Martin knew Earth lay. "Exploration revealed two worlds where civilizations had destroyed their own worlds before they could leave them." Flyover of a landscape cracked and burned. There were black bits of what might have been vegetation, and gigantic pieces of metal melted and splintered. "They have never before encountered a civilization like ours that is at a stage where it can destroy its own world and may do it before it becomes intelligent enough not to."

Something about the ruined planet, with great empty basins where oceans might have been, silenced Smith for an instant.

"OK, so what do these—what is the bottom line here?"

"We need to follow our own golden rule—celebrate our differences, not fight over them."

"Trust, but verify" whispered Smith. "Christophe, what do you think?"

"Do you mind?" he asked, and then without waiting for an answer, started playing with the keyboard. The computer shut down and began to reboot.

"You are a miracle worker as always, Christophe! What happened?"

"Not sure."

"Christophe, meet the gentlemen who created this catastrophe."

Christophe narrowed his eyes at both of them. "We have taken enough of the Chief's time. Come with me, gentlemen, and tell me how you did this."

"We didn't," said Martin, but they were swept into Christophe's wake by two Secret Service agents.

I/We recorded your voice. I can send this story to the entire planet.

Martin had two immediate thoughts—cool! And let's wait.

Why wait?

Like you, I am an observer of my own species. Let us see how my government reacts. You can always broadcast if Nathan and I are dropped into a dungeon.

Christophe led them into a room full of computers.

"Can I call my lab and check on them?"

One of the Secret Service agents shook her head.

I'll call them, Dad—I will tell them you are at the White House in a series of important meetings.

Excellent idea, Snehal.

I know they won't drop you into a dungeon, Dad, but might they put you somewhere, ah—

No dungeons, Patrick, but they might keep me in endless top secret meetings until I die of boredom. Glad you are here with me.

I/We are too.

"Please sit" Christophe indicated a chair in front of one of the computers.

"How about a cup of coffee?" Martin asked pleasantly.

"No—might spill on the equipment."

Tough interrogation—they are going to see what happens when I go into withdrawal.

Nathan snickered slightly as they put him in front of another computer.

"What's so funny?" Christophe asked.

Nathan gave Christophe his Buddha look. "You are wound pretty tight. Can't say I blame you. Take a deep breath or two. Our whole universe just expanded—we are not alone."

Christophe shook his head. "OK, whatever. I need you both to show me how you got into our system. Maybe you have collaborators elsewhere—if so, you need to contact them and tell them to reveal everything. I think you guys are probably all right—why else would you have showed us? But if there is a hole in our system, someone else could exploit it."

"OK," said Martin. "We are now testing a hypothesis—is there something else besides an alien intelligence that could explain the story you and the Chief of Staff just saw?"

Christophe nodded.

"Let me propose a first test. Nathan and I will allow ourselves to be scanned for any electronic gear—heck, we can do what you do in airport security lounges—then stand well back from the computers and this alien intelligence will show us—something. I am checking if that is possible."

Yes.

The Secret Service agents took them through a scanner after they had emptied their pockets. They both stood back away from the computers while Christophe and an assistant sat in front of them. The global cyberattack map popped up on the screen. Christophe started checking which web browser was open. Meanwhile the map zoomed in on a location—DC. Then a local map popped up, and they were zooming into the White House, with attack vertices coming toward it. The lines disappeared.

"Check your cybersecurity monitoring system—is anyone trying to penetrate now?" Martin asked.

Both monitors now showed Christophe looking at the computer. His phone rang. He picked it up automatically, staring at himself picking it up.

"Hello—"

"Hey, boss, got a message you wanted cyberattack status. No incoming at this point—they all stopped 30 seconds ago. What did you do?"

Christophe stared at the screen. "Boss?"

"Roger, thanks for the update."

He ended the call. Silence. "What else can you do?" Christophe asked.

I/We can reveal all online information about any of the beings you regard as separate and conscious on this planet.

"No!" Martin whispered.

"What?" said Christophe, turning toward him.

Tell the truth. "It said it could reveal all online information on any individual on the planet. I was telling it not to."

Christophe thought for a moment. "Tell the system to reveal my online information on just this computer—right after you all leave—and that when I hit the escape key, it should stop."

Martin nodded; he and Nathan stood up immediately and went out with the Secret Service agents.

Oh for a shot of coffee, though Martin.

I can feel your pleasure in this substance and begin to almost enjoy the rise it gives you. It has been too long since I have been in a body—

You can live in my body with me, but don't kick me out unless you can put me in a better model.

The door to Christophe's office opened and the Secret Service agents herded them in. He looked a little pale.

"You got things I didn't even know were online. You guys are a major risk—"

"We didn't get anything," said Nathan. "We wouldn't want to have access to anyone's private files."

Christophe shook his head.

"You two saw us," Nathan said, pointing to the Secret Service agents. "Did we touch anything?"

"Search them—head to toe. An MRI, too. Those guys are hiding something."

No! Martin could feel Patrick and Snehal's fear.

Interesting, Martin thought. *They have been connected all of this time and I don't have a headache.*

They have become more adept, and so have you.

Nathan was arguing with the Secret Service agents. "Are you arresting us? If so, we get a phone call."

"Not if I invoke a threat to the President," said Christophe. "What you did to me you could do to him, or to anyone."

Why is this man so—

We call it angry. Because you have displayed his personal information.

But only to him, and what does it matter?

"I need these two searched!" Christophe shouted. Other IT workers were looking up from their desks.

"Phone call. Lawyer." *Snehal, call Mom, tell her to call Simon.*

Who should I call, Dad?

No one—but stay with me.

"Hey—" One of the IT guys started punching keys, another went "Whoah—boss, you gotta look at this."

"Keep them here," muttered Christophe and he went to look over the shoulder of Mr. 'Whoah'. Martin could see his eyes widen.

Your planet, an island in space—

Spaceship Earth Martin thought back.

Hey Dad—we see Earth on our computer, too.

Christophe almost knocked Martin over running to his computer.

Enjoy the view, Martin thought, and heard his own words coming out of the speaker in a slow, somewhat distorted voice. He felt the hairs on his neck rise.

How—

I/We can translate the sound you carry in your mind into frequencies and put them through the computer.

Christophe was swearing. "Reboot the whole system!"

"Without backing up?" someone shouted from the IT bullpen.

"It's a virus, dammit—we have to get it off the network."

No virus—an alien intelligence lives in our global network Martin thought. He heard his own words echoed in the slow, somewhat distorted voice.

"Shut down now!" shouted Christophe. All the cell phones started ringing at once, including Martin's and Nathan's. The door opened. An intern stuck his head in.

"Hey, all the computers are showing Earth—looks like a space station view—Larry is about to blow a gasket."

"You two are in deep trouble," Christophe wagged a finger at them. His phone was still ringing. He opened it and gasped. Nathan opened his at the same time, looked at it and smiled.

"Earth—slowly turning."

Dad, Snehal and I can see it—we are passing over Central America.

Smith burst in.

"Christophe, the President needs to see you. Hey, you two follow us, the President may have questions"

You might get taken to our leader!

That is a joke from a story or stories you saw. Your jokes are one of your strengths.

Snehal, I hope you are telling your mother where I am and that we are OK.

Yes, Dad.

Patrick, tell Grandma not to worry.

A gofer was kind enough to get Martin a coffee while they waited in a conference room. Burnt—in the pot too long—but he felt the caffeine flood his brain.

Why is it so hard for you to concentrate?

Perhaps because we are in bodies? Perhaps because we can imagine multiple things at once? Perhaps because if we concentrated, we would not be able to hear wisdom whisper.

Smith came out. "The President wants to see you." The tone of his voice and his frown signaled that this was a bad idea.

Martin knew the Oval Office from the West Wing. It looked pretty much the same. He did not feel the awe many reported. The Oval Office was far from the center of the universe.

"Mr. President, this is Dr. Krishnamurti and this is Dr. Angell."

"Gentlemen." President Harnes was dressed impeccably as always, in a black suit. There was a gray dusting in his black curly hair. His eyes were hard, narrow. "I need you to stop projecting Earth over the entire planet and tell me how you managed to do it. I hope you are both patriots and will put this knowledge into service for your country."

The leader of the free world—though none of them were really free.

Can you restore the network to human control?

Done.

"I think you will find the networks are all back in human hands," Martin said. "But not because I control them—because I asked the alien—presence—to release them." Christophe looked at the screen on his cell, hit a key and started talking rapidly, turning away from the President and Martin.

"Christophe?" said the President, tilting his head.

"Oh—checking sir, to make sure we are up—and not just us."

"And—"

"It will take a while to be sure there is no damage. Those two—we need to figure out what they are doing."

"It is you that are doing," said Nathan. "Hanging desperately on to your familiar world despite evidence that everything has changed." Nathan smiled. "How about you, Mr. President?"

Amazing—Nathan looks like such a respectable straight arrow—

I am a child. Martin felt Nathan's laugh. *Trying to understand everything—and I like you because you are, too, though you take yourself so seriously!*

The President was looking at them, eyes slightly narrowed. "I am afraid I have to have the two of you searched, inspected—"

A PowerPoint slide showed the pattern of Voyager signals on succeeding days. Cameron Baird at the Deep Space Network was on the phone in the middle of the conference table. "The data is noisy because the signal is changing, but we are sure that Voyager is emitting a signal too complex for its hardware. We are checking with NASA to see what could account—"

"Mr. President." The PowerPoint had been replaced by a dark-haired woman with skin the color of teak.

"Chit Sung Ananda!" said Mohammed. Of course—the Myanmar dissident, under house arrest. Becky could breathe—for half an instant, her body had remembered Bin Laden—

"A divine message can come from many sources—from the face of a child, or an act of compassion, or a sunrise that reminds us what a miracle it is to have this light, this warmth, this wave of color spreading across the landscape. Now I wonder if a divine message can also come from another part of our galaxy. A being or beings have somehow interfaced with my consciousness and can speak without words. Just a few minutes ago, I connected with four other human consciousnesses in the office of the President of the United States—an astronomer, a psychologist, and their two boys, all of them connected with each other."

"Martin! And Nathan!" She hadn't meant to say it aloud. Heads half-turned for an instant, and then swiveled back to the screen.

"Many of you will immediately conclude I am out of my mind—and let me say you are not the first to think that." She laughed. "Because I believe we all have a right to freedom, to choose our own government, our own religion—and I would rather be imprisoned in my house than outside with an invisible muzzle and chains.

"This intelligence who communicates with us and connects us may be an opportunity to learn about ourselves. It wants to understand. It wonders why we chain each other."

Banging and shouting in the background.

"My minders have come." She put on the kind of patient smile you wear for a child who is acting up. The room filled with uniforms. A hand blocked the screen.

"Do you want me to go on?" Ralph's voice came from the conference phone in the middle of the table. Josh was on his cell already.

"Bottom line, Ralph. I need to get to the White House and see what kind of trouble my husband is making." Becky spoke while she was moving toward the door.

"Voyager has—"

"I'm coming, too" said Josh.

There was silence for an instant after Chit Sung Ananda disappeared. The Secret Service reacted first. "Come on, you two—"

"Belay that order, gentlemen," said the President. "I need to know everything these two know!"

"Excuse me." A short, dark-haired elemental force blasted between Martin and Nathan.

"Ah, Gabriella—perfect timing! Gentlemen, my National Security Advisor" Gabriella gave them a sharp once-over, then nodded. "Larry, ask State to find out what happened to Chit Sung Ananda and set up a Cabinet meeting soon as possible."

"Why am I doing this? What am I going to do when I get there?" Becky wondered as the Metro rattled under the Potomac. Aloud, she said to Josh.

"Nathan is level-headed, but Martin—God knows what he will say." *Or is doing to our son in this telepathic whatever.*

Josh ignored her—he was texting nonstop. As the train pulled into Foggy Bottom, her cell phone buzzed. Ralph.

"Hi—"

"You are heading to the White House, right?"

"With Josh Epstein."

"Who?"

"Secure and Trustworthy Cyberspace." Josh looked up, frowned, put a finger on his lips and looked around. *Oh my God—he thinks he is a spook because he handles projects that might help protect us against spookery.*

"I think I have figured out one place the outgoing signals are going." Ralph was excited "Voyager 1's return transmissions are sending us much more information about interstellar space than is possible, given what is on Voyager. Factoring in the delays in the signals to and from each spacecraft, the Deep Space Network calculates that the first hint of a change in the flow of information came after one of those outbound alien signals could have reached Voyager."

"This info may be worth reporting when I get where I am going in about 20 minutes—if I can get in. I'll call you back."

"Jim Clark was online at our meeting. Let me see if I can get him to escort you."

Martin and Nathan had been patted down, not wanded, and were asked to keep their hands where they could be seen. They were instructed by the President to tell the whole story, from the beginning. His secretary had placed a recorder on the table Gabriella was alternately studying them and the screen on a tablet.

"This is a deposition," said Nathan.

"That's right," said the President, "except there will be no legal action as a result, and this information will be shared only within the government."

"Which means it will be leaked to the *Post* by tomorrow morning" Martin said.

Gabriella frowned, but the President laughed. "You know Washington. Look, I need to understand how Chit Sung Ananda can be seen by us, and I presume by the rest of the world Christophe?"

"I have checked and it seems every computer and TV in the world saw her."

What happened to her? Martin thought.

A voice that sounded like Martin's came out of Gabriella's tablet. "Chit Sung Ananda searched by minders who stopped her. We are with her looking for chance to show all."

"How—" Gabriella looked at Martin and Nathan, with their hands on their laps. "That's not possible, this—these guys were searched, right?"

Her tablet said "How could two you see in front of you speak from your device? We can speak some language, help us do better."

"Christophe?" Gabriella looked at him, almost in panic. Christophe frowned, looked at the tablet, shrugged.

"Maybe it is aliens? I can't explain it."

Smith opened the door, and closed it discreetly behind him.

The President frowned. "What?"

"Jim Clark with two people from the NSF with news about Voyager One of them is his wife" pointing at Martin.

Mom! Said Patrick. I know you are safe now, Dad.

True—no one messes with your mom.

Becky had never been in the Oval Office. The President had his back to her, and next to him a dark-haired woman—must be the National Security Advisor. Martin stood up.

"Just greeting my wife—will keep my hands where you can see them."

Greet me—what—Before she could think he had his arms around her and turned her so the agent could see his hands. "Patrick says hi," he whispered. She shoved him away.

"Martin!"

The President laughed. "My wife doesn't like public displays of affection either. Doesn't stop me."

Martin touched her elbow. "Becky Olson, my wife and an NSF Program Director in astronomy, Sir." He, of course, remembered only part of her title.

The President stood, shook her hand, nodded to Jim Clark, who introduced.

"Josh Epstein, Secure and Trustworthy Cyberspace, Sir." Another handshake, and the Secret Service agents brought up more chairs.

"Larry, no more interruptions, please!"

"Cabinet in 30 minutes, Sir." He closed the door.

Your greetings are slow.

Becky started. Martin's voice, coming out of the laptop. *Perhaps that is good. I do want to understand why you spend time—chaining?—each other and also can—love.*

Is that the right word?

Yes, because true love is not a chain As Martin said it, he saw the truth of it.

True love is not a chain, his voice said out of the laptop. He felt Becky start again, and put a hand on her thigh. She did not remove it.

Millions of years ago three species that became us fought each other.

"And now you are peaceful and enlightened but have taken over our cyberspace and are doing something to Voyager." The President steepled his head on his hands and frowned at the laptop.

I/We— I am also we—I am in touch with home planet and here at same time—cannot tell you how—emerged from a signal that entered your simple network. Can show you home planet. Behind the signal was waves of particles,—you have only wave and particle, which reflect who you are, not what is beyond you. These wave and particle have been called to Voyager and are using it to introduce you to any of your neighbors who might encounter—

Neighbors!

The door opened. "Yuri Kostov on line 1, Sir."

"Hold him—"

"President Harnes."

The phone was on speaker. President Harnes grabbed the receiver.

"What can I do for you, President Kostov?" The speaker was still on. Harnes waved to Christophe, who tried punching buttons.

"You took control of our Internet earlier today. You must reveal how you did this, and promise in writing never to do it again—"

"Yuri, we had no more control over our cyberspace than you over yours—"

"Nonsense. This is an act tantamount to war—"

"Our systems have been compromised, too," said President Harnes. "And I think we have found the perpetrator, though I can scarcely believe it."

I/We are your IT systems, now. I/We will not interfere with your work; I/we will try to understand it, and make some improvements to enhance communications.

"You must think I am a child or fool. Goodbye, Mr. President!" They all heard the click—but Yuri was still on.

"You must talk to this President Every signal sent to your—"

What is the word for those who carry crude weapons and kill others?

I think you mean soldiers. They also have missiles. Martin tried to conjur up an ICBM.

Soldiers, and to the missiles you would send against others can be sent to everyone.

"What—you dare to threaten me?"

"Not me, Yuri—whatever this—thing—is that is in our IT system."

"You are making fun of me, of us. We will not stand for it," Yuri growled and Gabriella's screen switched to part of a face shouting in Russian.

"Dammit, you have to tell your alien friend not to interfere with diplomacy!"

I/We are putting everything the Russians say on your Internet where everyone can find it. All of what the—minders—of Chit Sung Ananda say. You know that some of what you do is evil—your word, your concept. Evil will no longer be hidden.

Chapter 5

Manipulations at the Nanoscale

Mark Wiesner and Hélène Crié-Wiesner

Professor Jill Kruse had a slight hangover. Her retirement celebration with colleagues in the engineering department the day before had been a bit too festive. "I would have been better off sticking with red wine," she grumbled, rubbing her head. "Too much champagne doesn't mix well with a 75-year old." She glanced around the office she was about to desert in June 2030. Her diplomas, awards, file cabinets stuffed with academic works, countless memories accumulated during an academic career spanning half a century . . .

Jill mindlessly dusted an old photograph with faded colors, half hidden by the screen of her computer and holographic imaging set. She appeared with her students in white coats, holding sampling gear in front of a sort of oversized planter containing lush vegetation. "How old was I?" she murmured, turning the frame, moved by the memory of these mesocosms, where for a decade, she had scrutinized the environmental behavior of various nanoparticles. "It must have been about 2010, just before Congress cut funding for research on nanotechnology! Hmm . . ."

Flash Forward: A Series of Futuristic Vignettes
Edited by Nora Savage and Anita Street

ISBN 978-981-4669-44-3 (Hardcover), 978-981-4669-45-0 (eBook)
www.panstanford.com

The old lady sighed—no use in stirring up unpleasant memories. Granted, she had lost almost 10 valuable years working in a scientific field that went nowhere—in the United States, at least, where political calculation and a chill in public trust of science had prevailed over reason. But the rest of the world, Europe, China, China, and China . . ., moved on, pursuing a trajectory of nanotechnology development that soon provided a steady stream of sophisticated products now flooding the US market. Nauseated, preferring not to dwell on the past, Jill flipped the forgotten photo into the wastebasket.

The screen of her Chinese-made office telecom unit had been blinking for a while, when the professor finally decided to take the call. She would have liked to continue alone with her thoughts, drifting along with her nostalgia. Retirement, even if you are a recognized landmark in your field and even if your status as Emeritus Professor ensures that you will not just drop out of circulation, means you're old. Jill didn't like that idea. The ID displayed on the screen indicated a call from the National Science Foundation (NSF), which piqued Jill's interest. "Maybe the NSF needs to pull an aging scientific icon away from retirement to rehabilitate their public image," she thought, laughing to herself as she placed her finger on the Chinese print recognition device and answered the call.

The despondent face that materialized on the screen instantly dashed any fantasies of the professor.

"Jill, I thought you would want to know. They've finally got what they've been asking for. Congress has zeroed out our budget. The NSF is all but gone."

"Wait, Eva, I don't understand, I thought that the committee was leaning our way?"

The director of the NSF was seated with her back to the window so that Jill could only make out the silhouette of her old friend against the backlight, but Jill could sense the emotion in her voice. The woman who had headed the NSF during the last four tumultuous years took a deep breath.

"It's not official yet. The committee suspended its session before moving to a vote. But Senator Hatchette just called me to give me the news: it's a done deal; Congress will finally get what it's been after—the NSF's budget for research will be pushed to pay for so-called private sector research. Jill, you know it's all theater. Even industry supported us, but between telling it like it is with climate change, all this talk about trimming government, and then the nanotechnology scandal—well, that's all they needed to find another 14 billion to spread around to some more deserving characters," Eva scoffed.

"But can't we shut this down in Congress?"

"You know as well as I do, Jill, that the committee wouldn't go out on a limb if it didn't have the votes. It's the same script it followed when it eliminated the EPA and the Department of Energy last election cycle. They're just cleaning up loose ends now. These guys are allergic to government and hate science when it gets in the way of their business plan."

This was not the Eva that Jill knew. Politically charged commentary like this was completely foreign to this well-groomed academic, renowned as much for her diplomatic skills as for her scientific knowledge. With the end of the NSF, and near retirement herself, *what did Eva have to lose?* Jill thought, as she considered the irony that voice recognition algorithms developed years ago at the NSF were no doubt at work digitizing their every word for storage at some remote National Security Agency facility.

"It's been uphill every day since nanogate, hasn't it?" she squeaked.

"Uphill? Yeah, that's putting it lightly. When *The Washington Post* uncovers a case of twenty-first-century Lysenkoism and research misconduct at that level, it takes down the entire credibility of U.S.-sponsored science. When I think that Victor sat where I am now—it's no wonder that Congress wants the NSF's head on a platter. What's amazing is that we survived this long"

Jill rummaged under the table, retrieved the old mesocosm picture, and stuck it in front of the camera.

"Do you remember? Before all this mess started?"

* * *

Twenty-five years earlier, before a reshuffling of the geopolitical landscape, economists still spoke of "peak oil." Now in 2030, Middle Eastern reserves are inaccessible in the midst of a string of escalating crises that make it impossible to get oil out of the Persian Gulf. But Middle Eastern oil is a concern of the last millennium. The energy crisis, at least as it predicted since the 1970s, had never materialized. The United States tapped shale gas reserves, imported Canadian tar sand oil, and mined coal—largely to sell to China. In the process, mountains were leveled and valleys filled across Appalachia. Regulatory barriers on oil and gas drilling offshore have been more or less eliminated along with the Environmental Protection Agency (EPA) Europe, juggling France's aging nuclear inventory, Germany's growing wind and solar farms, Eastern European coal reserves, and Russian gas, continues to muddle through an extended period of economic stagnation that has pushed the political spectrum across the continent to the right.

But while US reserves of gas and oil are flowing, and Russia taps into its immense gas reserves, the effects of global warming are accelerating. China had no qualms about burning foreign coal and expanding its nuclear capabilities 20 years earlier. But through its dazzling innovations in nanotechnology and its reserves of scarce minerals needed to make these technologies, China is now the world's primary supplier of clean energy in the form of photovoltaic and energy storage. China's technological dominance took the United States by surprise after American progress in the field became hopelessly lost in an ill-conceived strategy based on bad data and corrupt leadership.

Early US research suggested that nanomaterials made of several rare-earth metals showed extremely high toxicity at the nanoscale. Public backlash was immediate, with calls to avoid another genetic engineering or nuclear energy fiasco. Sensing a string of lawsuits that would dwarf those from asbestos, and the Fukushima disaster combined, investors panicked in a short-term selling frenzy of mining interests.

That created an opening for Chinese investment in strategic mining interests at rock-bottom prices. Curiously, over the next few years, these investments were followed by a string of announcements of major scientific discoveries in nanomaterial-based electrical storage, photovoltaics, and—the holy grail of energy production—inertial confinement fusion. Needless to say, the price of mining shares rapidly recovered. Now, in addition to their own vast reserves, China and its international commercial interests control much of the world's supply of a list of exotic elements with names like samarium, ytterbium, and neodymium, coming from mines in India, Brazil, Africa, Australia, Canada, and even the United States. The entire operation has the stink of insider trading. Calls for the US Congress to investigate seemed to go nowhere, as anonymous donors with questionable foreign ties continued to sweep boulder-sized crumbs from the table of high finance into campaign coffers of sympathetic candidates.

And Congress took the bait. When Congress finally put an end to the EPA in 2020, it had its sights fixed on increasing fossil fuel production. It used the old and familiar argument that removing regulatory barriers to free enterprise was the fastest path to allowing good corporate citizenship to bloom. Environmental and economic prosperity would follow. And in the short term, energy prices dropped, taking some pressure off the economy.

But by 2025 climate change was undeniable—even to the previous nonbelievers for whom arithmetic was an article of faith. It was apparent that sea levels were rising faster than even the most pessimistic EPA models had predicted. In the United States, three particularly violent hurricane seasons devastated the cities of Miami, Mobile, Houston, and New Orleans. Droughts in the western United States and early winter storms associated with polar vortex incursions on the East Coast took taking a devastating toll on US agriculture.

The turnaround in political posturing was stunning. In the 2028 US presidential campaign, former climate change deniers claimed that they had seen global warming coming all along. Reading from the same script they had used to eliminate the EPA, the responsibility for hurricanes, floods, and early freezes was placed squarely on the back of an inefficient public sector, government meddling with private sector innovation, and faulty government-funded models that had incorrectly forecast much lower levels of global warming.

Politics alone could not have done the job of putting the NSF on the same path toward extinction that the EPA had taken 10 years earlier. If the average American is going to have a hard time sorting out the good science from the bad, it is because scientists themselves muddied the water and gave their adversaries the ammunition they needed to discredit their work.

* * *

As Jill ended the teleconference with Eva, she again looked at the picture and returned to reminiscing about the role she had played in this national tragedy. There had been good guys and bad guys, but initially they had all been on the same side: the side of science.

* * *

At the beginning of the millennium, at 45 years of age, Jill's career had been in high gear. A tenured full professor at the Texas Institute of Science and Technology (TIST), Jill chaired the Department of Environmental Science and Engineering at TIST. As a child growing up in Oklahoma City, she never would have dared imagine that she would one day serve on the faculty of this distinguished private university, sitting in faculty meetings with some of the most prominent names in high-tech start-ups, Nobel Laureates, and members of the National Academy. In fact, being admitted to study at a university of this caliber seemed implausible, let alone teaching there. There had been no engineers in her family and role models for women in science and engineering were rare.

When she submitted her applications for university study in chemical engineering in the early 1970s, with good grades in all the wrong classes, she ultimately landed a full scholarship with in-state tuition at the University of Oklahoma. But it was environmental issues that ultimately captured her imagination. She helped organize Earth Day activities with a few like-minded students on campus and planned on going to work someday with the newly created EPA. After finishing her studies in chemical engineering, she shifting her studies to environmental engineering as a doctoral student at Cornell University and was promptly snatched up after graduation by TIST in its newly created Department of Environmental Science and Engineering. Jill had spent the last decade at TIST, building a dynamic, multinational research group of doctoral students and postdocs working in the area of water purification.

Jill recalled eating lunch at the university's faculty club one day in October 2000, where she had arranged a meeting with one of her most celebrated colleagues, Robert Cortely.

Cortely, a physicist and member of the National Academy, was rumored to be a perennial candidate for a Nobel Prize for his work on the use rare-earth nanostructures to create high-energy lasers for inertial confinement fusion. Cortely had established TIST's Center for Nanoscale Science and Energy Technologies (CNSET). "Sunset," as the center was called, had as its mission the development of nanotechnologies for energy applications and included a portfolio of projects aimed at advancing alternative energy technologies including solar, fusion, transmission, and storage.

Cortely was working on a proposal to the NSF to obtain major funding for Sunset research. Jill and her students were occasional users of the CNSET's impressive facilities that included advanced nanoscale imaging capabilities and Jill had suggested the meeting to discuss how they might work together on the NSF application. That day, Cortely was venting his frustration at his salad

"Frankly, Jill, trying to get NSF funding to study the potential risks of nanomaterials on the environment doesn't seem to me to be exactly the idea of the century. We're just starting to make progress putting the National Nanotechnology Initiative in place, and this could really throw a wet blanket on everything we are trying to build. Anyway, the whole idea seems premature! We have no idea if these materials will cause a problem."

Jill could see Cortely's point. Just last year, Cortely's colleague, Nobel Laureate Richard Smalley, had captivated members of the US Congress, explaining, "We are about to be able to build things that work on the smallest possible length scales, atom by atom, with the ultimate level of finesse. These little nanothings, and the technology that assembles and manipulates them—nanotechnology—will revolutionize our industries and our lives." Months later, President Clinton had launched the National Nanotechnology Initiative (NNI) in his budget request to Congress. Cortely's vision for energy and nanotechnology depended on the mobilization of resources at the national and international scales and long-term political support.

Closer to home, TIST was planning a major fund-raising effort and Cortely was the centerpiece. A new building, instrumentation, new faculty positions . . . with resources like these Cortely could make real progress. Raising doubts about the safety of nanotechnology might scare off donors.

But these were exactly the reasons that Jill had proposed the idea to Cortely. "That's the whole point, Bob. If we wait until it's a problem, it's too late. This time science needs to be out in front. You know as well as I do that it's just a matter of time before people will accuse you of playing with fire." Nanorobots and all . . .

"Yeah," Cortely cut in with a little impatience, "I've heard these delusional science fiction fans. It's hard to believe that some of those people call themselves scientists . . ."

"Rob, you know better than I do that there are always a few people looking to sensationalize on fear. So let's nip it in the bud. In any event, we're not allowed to ignore the question any more. We've lived through DDT, asbestos, and Love Canal. That's why we have the EPA. Investors want answers to these questions as much as anyone else. Let's give people answers before they ask the questions, and take a look at what these new materials do in the environment. Let's see if they pose a risk to health and the environment or not. In the long run we'll come out on top."

Cortely shrugged. He could relate to the idealism of his younger colleague, but he still thought that the case was risky in terms of public relations. But he knew that Jill was right. *If the CNSET doesn't take the lead, some other university team will raise the question eventually.* Weighing the pros and cons, he decided that building a proposal around the theme of environmentally friendly energy technologies, there might be just as much chance that the environmental theme would help sell the proposal as sink it. Better to be pioneers. He rose from the table.

"OK, we'll keep your project in the mix and hope that it doesn't torpedo the whole center proposal to the NSF. But you are going to need people who are as good at politics as science to pull this off. Expect to experience some unfriendly pressure."

And so the idealistic, if not naive, Professor Kruse, specialist in applying nanomaterials to create new water treatment technologies, began to assemble a team of environmental researchers who would be part of Curtely's proposal to fund the CNSET to develop green energy technologies.

Preparing a research proposal for the NSF was never an easy task. Normally, the chance of getting funded was less than 10%. Competition for center funding would be even more difficult, with virtually every top university striving to obtain a new nanoscience and engineering center (NSEC). NSECs provided longer-term funding than the smaller individual grants and Jill wanted to do something transformative at TIST—something that would bring together nanochemists, environmental engineers, ecologists, and ecotoxicologists to work directly with Cortely's group developing energy technologies.

When he came into the project at Cortely's suggestion, the young chemist, Victor Hobbs, dazzled the group with his hard work and dynamism. Jill was both impressed and a bit taken aback by the take-charge attitude of this young assistant professor. But Jill was pleased to work with Victor and the two became inseparable during the proposal-writing process.

Victor had come to TIST from Stanford and had succeeded in obtaining a string of prestigious young investigator awards on various topics. Victor seemed to be struggling to find a research theme to build his group around, but Cortely and Jill immediately saw the advantage of having the diligent Victor on the team.

As chair of her department, Jill seemed to have an endless line of people outside her office, urgently needing to speak to her about salary raises, space for laboratories, or, more mundanely, conflicts between office workers, accusations of student cheating, backed-up toilets, and parking. It was clear that Jill would need additional help if the center were to function as she envisioned. Environmental research was completely foreign to Victor, and like Cortely, he initially challenged the idea of studying the environmental impacts of nanomaterials. But when Jill suggested that Victor might serve as coleader of the environmental thrust, in addition to the role of nanochemistry leader that Cortely had already proposed, Victor put aside his reservations and jumped at the chance. Jill and Cortely walked Victor through the proposal process, and by the time it was written, all the boxes were checked—international participation, industrial collaboration, undergraduate education, outreach, diversity . . . and very good science.

The proposal selection committee was mesmerized by the TIST presentation led by the venerable Professor Cortely, who provided the final summary: "Clean technologies for generating energy will not only light homes and hospitals around the world. Nanotechnology will allow us to tap into the oceans to provide clean water that will support agriculture and improve sanitation. Water is the next oil. Copying nature and building from the smallest scale, we will have the capability to manipulate and assemble matter to provide water and new energy technologies that will revolutionize our industries and our lives without harming the environment."

In 2001, the team officially started its work on the rebaptized Center for Nanoscience and Research in Energy Sustainability (CNRiES). The acronym was a stretch, but the development office of TIST was much happier raising funds for a group with the name "Sunrise" rather than "Sunset." So when Victor proposed the name change, every one congratulated him on this brilliant stroke of public relations.

Indeed, Victor proved to be the ideal communicator to promote the centre's research. Victor was the sort of scientist who loved journalists. Eager to give interviews and able to distill complex concepts into sound bites that resonated with the general public, he left his nonscientist interlocutors with the feeling that they actually understood the entire breadth of the science performed in the university laboratories. Above all, Victor came to understand that stories on brilliant advances in energy technologies are of less interest to the press than stories describing a new potential threat to human health.

This became evident after Jill presented an early summary of Sunrise research to the EPA, entitled "Nanomaterials, Friends or Foes?," in which she presented the advances in energy technologies coming out of Sunrise and the need to verify the environmental sustainability of these technologies.

As Cortely had predicted, Jill soon found herself embroiled in controversy. Picked up by *The New York Times* among other international papers, the press focused on Jill's speculation that nanomaterials might be dangerous, largely ignoring her nuanced insistence that reliable data resolving the question of nanomaterial safety would require time to produce. It made for a much better story to label nanotechnology the "new DDT." An impressive array of new energy technologies coming out of TIST was in no position to compete with a threat like that.

Feeling pressure from donors who had been anxious to give millions to TIST's nanotechnology initiative, the university administration of TIST met to discuss the "Kruse problem." It was decided that the dean of Engineering and Applied Sciences, Karen Barrows, would deliver the news.

"Jill, you're just too close to the issue and a little too passionate about the subject. You need to step away from the leadership in Sunrise and let Victor deal with the press. Of course you can still do your research, but we need to take the heat off Sunrise."

"Passionate? Karen, I expected more support from you of all people."

Karen knew better than most what challenges awaited a woman trying to advance in engineering. As an early mentor, she had done everything in her power to make sure that Jill was on an equal footing with her colleagues. "Passionate" was the word that the provost had used in their discussions. It had sounded to Karen at the time like the sort of veiled insult tailored to discredit a woman. Karen was ashamed to have the task of reining in her protégé. But unable to speak in her own voice, she had merely parroted the provost's words. Now she felt even worse.

"You're right, Jill. I'm sorry. That's completely off base. But right or wrong, that's the way some people here are describing the situation. We're not going to change that. What's important is that your research continues."

Jill had been a bit embarrassed by Victor's propensity toward exhibitionism in the media and a tendency to sensationalize, but the bottom line was that he was doing his job. Even if his research was not setting the world on fire, his publications on nanochemistry coming out of Sunrise were solid. He had taken on the bulk of the work organizing internal meetings and scientific conferences; this allowed Jill to focus on the environmental research in Sunrise.

So it was that Jill acquiesced and slipped into the background, making Victor the public face of environmental research in Sunrise.

In the meantime, 2003 was a great year for fans of fright. Michael Crichton's book *Prey* was published, describing a potential takeover of society by swarming nanobots. A Canadian NGO published a report on the dramatic potential dangers of nanotechnology, which caused a stir in Europe and the United States: *From Genomes to Atoms, The Big Down*. This report presented an apocalyptic picture of what to expect if a ban on all technologies based on nanomaterials was not immediately put in place: "... control of nanoscale matter is the control of nature's elements (the atoms and molecules that are the building blocks of everything). Biotech (the manipulation of genes), informatics (the electronic management of information), cognitive sciences (the exploration and manipulation of the mind), and nanotech (the manipulation of elements) will converge to transform both living and nonliving matter. When genetically modified organisms (GMOs) meet atomically modified matter, life and living will never be the same."

Other "public interest groups" joined the call for a nanotechnology moratorium, in particular the Houston-based group Allies of Creation, founded by Peter Worldish. Allies of Creation presented itself as a faith-based group dedicated to "protecting life and the divine natural heritage in the face of technology." Having started as a group opposed to all forms of medical intervention in human reproduction, including contraception and in vitro fertilization, it had expanded its portfolio to include resistance to GMOs, geoengineering, and neuroinformatics.

Worldish's activities seemed better formulated to advance a sensationalist agenda designed for self-promotion (and to ensure a steady flow of money from the pockets of the Allies' donor base into the coffers of his organization) than an honest exploration of the consequences of new technologies.

However little credibility Jill, Cortely, and Victor initially attached to Worldish and his organization, Victor was obliged to engage Sunrise in a public dialogue to respond to the Allies of Creation's alarmist incantations that increasingly reverberated among the general public. As Victor and Worldish increasing crossed paths,

Victor began to see a path forward for his own research program and his career. Fallout from the "Kruse problem" at TIST had faded as the public dialogue on nanotechnology had expanded. It was clear now that TIST did not control the public debate on new technologies. The administration and donors alike agreed that positioning TIST as a leader in addressing the risks of nanotechnology was good business after all.

Victor would subtly reposition the original theme of *environmentally responsible technologies* in Sunrise to one of *environmental threat*, bringing in more money and more attention. In the process, Victor and Worldish would become partners rather than adversaries.

For Jill, academia had the advantage of providing a stable, if not sometimes isolated environment, where "truth" in the form of scientific certainty could be pursued without regard to politics and media outcry. Facts would necessarily prevail over ideological ramblings. And in fact, at first, a strategy of denial worked perfectly well for Sunrise, as well as countless US research labs working in the field of nanotechnology: a moratorium on nanotechnology? What a ridiculous idea! If one country decided to freeze its research and industry in this area, other nations, other scientists, and other industrial interests would move into the void. The train had left the station.

Moreover, the first results of research on the behavior of nanomaterials in the environment began to arrive in Jill's laboratory—nothing to report. Both laboratory experiments and long-term field experiments performed in TIST's mesocosms failed to yield results that seemed the least bit alarming. This was good news for nanotechnology but bad news for continued funding.

But Victor's recent forays into the impacts of neodymium nanoparticles' toxicity went against the grain. "More toxic that strychnine" read the draft manuscript describing the most recent work coming out of Victors' lab.

"Victor, this just doesn't make sense." Jill had done similar experiments a year earlier and the mechanism that Victor proposed for the toxicity of the neodymium nanoparticles ran contrary to Jill's results. "Are you sure it wasn't the solvent that you used in making these particles?"

Victor had been trained as a chemist, not a toxicologist or environmental scientist. His recent forays into toxicology seemed to Jill to be just one more dalliance in his quest for a research identity. Lacking the creativity for fundamental discovery, Victor had built his research program by appropriating topics at one time or another from the Cortely or Kruse labs as a basis for writing new proposals or publishing minor papers. Now, he was collaborating with a young cell biologist at TIST who certainly had the capability of producing reliable data.

But Jill had her doubts about how closely Victor's work was being followed. Studies of this sort were not the sort of thing that you just learned in between snacks.

Victor gave Jill a dismissive grunt. "We used a control. The solvent showed no effect. I suggest you go back and look at your own results." Victor had been granted tenure the previous year and had taken on an increasingly defiant stance with his colleagues.

"Victor, you know that Cortely is livid. These particles are at the heart of the laser confinement technology he has been developing. If your results hold up, then they need to be published, but the whole story needs to be cross-checked before publication. One of us is wrong and we need to get to the root of it."

"I've checked them enough for my satisfaction. Cortely's going to have to deal with it. This is going to be an important publication."

Jill had no doubt that if published, the paper would clearly have a major impact on the field, for better or worse.

In retrospect, Jill criticized herself for remaining silent in the interest of collegial harmony and a small doubt about whether these particles really were toxic. After all, wasn't it better to err on the side of precaution? But later it would become clear that Victor had, perhaps at first innocently, crossed the line into a case of research misconduct. Maybe Victor knew that Jill had been right all along, or maybe he felt that the data just had to support such a logical theory for the toxicity of these rare-earth nanoparticles.

In any event, when the paper was published in 2005, the Peter Worldishes held the paper up as proof that the fear they had been peddling was justified. With difficulty publishing negative results showing that one nanoparticle after another was not toxic, and relatively little interest in the papers that were published, Jill's research funding stagnated.

Victor, in contrast, had been approached by numerous industrial interests eager to fund his research. The identity of the industrial donors was often obscure and would later be traced back to a few key Chinese investors. But the short-term result was a massive injection of research dollars into Victor's lab and a happy university administration.

It was just after the formation of Victor's Industrial Nanotoxicity Consortium (INC) that Cortely unexpectedly died of a massive coronary attack, leaving Victor as the heir apparent for leadership of Sunrise. This was ironic, given that in the year preceding Cortely's death, Cortely and Victor were no longer on speaking terms. Cortely shared Jill's skepticism of Victor's motivations, if not his data.

The personal interests of Victor and Worldish, the tireless crusader for a moratorium on nanotechnology, had begun to converge. Worldish needed scientific backing to confirm that nanomaterials were dangerous, meriting a national moratorium on their production. Worldish's crusading only served to place Victor, the first specialist to have "proved" the danger of nanomaterials, in the spotlight. Victor was becoming an academic and media superstar, climbing the academic ladder and amassing larger research budgets.

By 2006, Worldish and Victor were meeting in televised "citizen debates," sharing the podium at international colloquia, and even providing back-to-back testimony as Congress considered research funding for nanotechnology.

Bolstered by massive funding from the INC, a startling series of papers came out of the Victor's lab from 2006 to 2012, confirming the toxicity of an entire series of rare-earth metals. When attempts by others to reproduce the results failed, researchers contacted Victor, who explained the complex procedures involved in preparing nanomaterials. While most of these researchers were experienced toxicologists, they were unfamiliar with, if not intimidated by, the new area of nanotechnology. Victor would usually propose that they collaborate, providing them with the same nanomaterials used in his experiments. They used the materials provided by Victor, under the assumption that their nanochemist had followed all the quality assurance procedures that were at least equal to those used in their own field, and in doing so, they replicated his errors.

Jill's work continued at the methodical pace that she felt was required to conduct good science. In the interest of Sunrise, she was required to maintain a professional stance with Victor. But the tension between the two of them had only grown following Cortely's death.

As the leading spokesperson for Sunrise, Victor had the habit of presenting his late colleague's work and the work of others in Sunrise without attribution. The words "I," "me," and "my," tended to dominate in Victor's presentations, moderated only by a strategically placed royal "we," leaving the audience to assume that all he presented was the product of Victor's brilliant intellect. Jill was just as happy that recently Victor seemed to consider her own work as inconsequential.

But Jill looked genuinely happy in the 2010 photo of her standing with her students amongst the thick foliage of the mesocosms. They all understood the urgency of producing good data. The group had become a tight-knit family working with the knowledge that the long hours spent in lab and field would make a difference in the trajectory of an exciting new technology. The effects they had observed so far indicated interesting, if only subtle, influences of nanoparticles on the mesocosms. The small amounts of toxicity they had observed in the lab seemed to be tempered over time in the more complex environment of the mesocosms.

It was in 2010 that the Supreme Court of the United States came out with decisions in *Citizens United vs. Federal Election Commission* and the Speechnow.org cases, which opened the floodgates for contributions to political action committees. Some of the same obscure sources funding Victor's research were ultimately linked to even larger contributions to more than a dozen political action committees.

Worldish never did obtain the nanomaterial moratorium that would have effectively brought an end to his career as a nanoactivist. But the combination of fears of rare-earth nanoparticle toxicity and the mere rumor of a moratorium were sufficient to depress the stocks of many nanotechnology start-ups and established mining companies. There was an uptick in contributions to Allies of Creation when, just before his death, Cortely had demonstrated enough progress in nanotechnology-based energy technologies to show that at full scale, they could outcompete the conventional fossil-fuel-based industry.

The Washington Post article that ultimately exposed the "nanogate" scandal detailed the role of politically sympathetic NGOs. It singled out Worldish with his ties to the Houston oil and gas community in particular, which undermined public support for alternative energies. Price supports for these new technologies were eliminated and the proposed carbon tax on fossil fuels was never implemented. A short-lived government loan program provided financing for alternative energy start-ups to scale up results from the laboratory. But for every 10 successful alternative energy start-ups that went unnoticed, Worldish was able to find one case where he could twist facts to paint a picture of misuse of governments funds, casting enough doubt to eliminate the entire government program.

Chanting the mantra "drill baby drill," the country leaped forward with fossil fuel production and undercut the price of the newer alternatives. In the meantime, even if Cortely's technologies couldn't find the investment support from a skittish venture capital market in the United States, they would soon be reproduced and developed abroad under the protection of more visionary political agendas.

By 2012, NSF funding for Sunrise had come to an end and Victor had been named dean of engineering and applied sciences at TIST. As director of Sunrise, Victor had established an impressive list of contacts, which included researchers and executives at many of Houston's energy companies. Victor was valued for his "down to earth" views on energy technologies. After all, his research showing the toxicity of the materials used to make the very technologies that "his" research center was developing testified to his pragmatic, objective character. And it didn't hurt the oil business. As dean Victor then turned his attention to working with Texas energy companies to develop "green oil and gas" research at TIST.

Victor's tenure as dean was not noteworthy. Chosen over the objections of the faculty search committee, he was the administration's choice. As dean, he did not raise money for a new building. But he maintained a veneer of success by creating three endowed chairs and "guiding" the school through the tail of a major recession that had seen TITS's endowment drop by 50%. He balanced the budget by cutting department staff, eliminating university support for graduate students, and furloughing faculty.

Presenting his time as dean at TIST as a major success, building on an international reputation developed from his nanotoxicity work, showered with glowing recommendations from TIST administrators, and, above all, having an unfailing ability to talk the talk, he was able to land the job as provost of Stanford. His departure from TIST was celebrated only after he had moved to California in the summer of 2015 and it was clear that he would not be coming back.

During Victor's short term as provost at Stanford, he was able to capitalize on his Houston connections to a much greater degree than he had been able to at TIST. He raised funds for several major construction projects and successfully endowed his new Green Carbon Institute with donations from individuals and energy interests around the world. Peppered with words like "responsible," "sustainable," and "security," the institute's mission statement was to "explore technologies and policies for minimizing the environmental footprint of fossil fuels, while adapting to climate change." Political connections he had made as dean at TIST multiplied at Stanford. When the 2016 elections ushered in an overwhelming change in Congress and the White House, Victor could count many of the Washington newcomers as close associates.

It was no surprise then when Victor was tapped in 2019 to serve as director of the NSF. The new crowd in Washington was cleaning house and Victor was just the man to help them do it. The EPA had become a perpetual thorn in the side of the oil and gas industry. Many politicians had campaigned on the platform of reducing government intervention, starting with an elimination of the EPA.

During the 2018 midterm elections, senatorial candidate Warren Hatchette, who would serve on the Senate Commerce, Science, and Transportation Committee, summarized his party's views:

"We are the party of energy and enterprise. Our party stands for a future where government gets out of the way of individual initiative and ingenuity and enables people to do their best, instead of telling people what is best. We demand an end to intrusive government meddling, an end to the high taxes on businesses, and a dramatic downsizing of government. No government is good government."

The path to downsizing government involved selective elimination of those pockets of resistance in the government bureaucracy that were impediments to gas and oil revenues, such as the EPA. And by 2020, the EPA existed no more. Where entire agencies couldn't be simply eliminated, budgets were realigned to more politically supportive sectors. Like-minded individuals were needed at the heads of organizations—the National Oceanic and Atmospheric Administration (NOAA), the NSF, and NASA—to shift funds from climate change research to other functions. For example, NOAA's climate change money was used to outsource data collection and to support "outreach efforts" in subsidies to media conglomerates.

During his six years as director of the NSF, Victor applied some of the same skills he had learned as dean at TIST. He balanced a shrinking budget by eliminating research on climate change and environmental protection or by reducing the budgets for ecosystem biology and environmental engineering. The number of program officers was reduced, and the size and number of grants awarded were cut back. He gave instructions that the remaining research dollars be reserved for groundbreaking research rather than visiting previously studied topics. Naturally, new studies looking at the environmental impacts of nanomaterials, once in vogue a decade earlier, were no longer funded. So Victor appeared to have covered his tracks. Unfortunately for Victor, as at TIST, he left a trail of unhappy and disgruntled people.

Jill never knew exactly how it came out. She had moved on to a life after nano, working on microbial-based electrochemical means of generating electricity from waste. But someone, or more likely several someones, alerted a certain senior Washington journalist to a pattern of conflicting data in the peer-reviewed literature.

What had previously been dismissed as a domestic squabble between scientists regarding the toxicity of several obscure nanomaterials took on greater proportions when the scientist who had "proved" the danger of nanomaterial was now director of the NSF.

And the squabble took on entirely larger proportions when Jill's work, ignored for the last 15 years, was corroborated by the more recent Chinese work, showing the rare-earth nanomaterials to be largely benign.

These developments took on an entirely new meaning when foreign research groups revealed that they had successfully scaled up revolutionary new energy technologies, largely following the plan outlined in Cortely's research publications. Naturally, these technologies were enabled by the very nanomaterials that Victor had "proved" to be toxic.

And still yet, this might have remained beyond the notice of all but a few scientists and bureaucrats, because no one had the power to open the books and follow the flow of money to reveal this critical fact: the control of most of the mining interests required to produce these materials was now held by a few offshore companies and foreign governments. The headwaters of that river of money extended to tributaries that financed research to "prove" the toxicity of nanomaterials and flowed to lobbying interest that managed to slow the development of new energy technologies in the United States. No one could wade into security protocols and legal protections that concealed this swamp of interconnected financial transactions.

It took a hurricane.

Was it a disgruntled employee at NOAA who chose the name for the early-November hurricane that hit Houston in 2025? The storm surge from Hurricane Victor was over 30 feet when it made landfall. Two days of record rainfall preceding the hurricane resulted in massive flooding as the storm surge marched inland, knocking out power grids and inundating most of central Houston.

Perhaps it was a coincidence or a well-informed act of revenge on the part of someone in the office of research accounting who also worked in accounting at Allies of Creation. But in the massive work following Hurricane Victor to restore damaged computer files and reconcile financial data from secure backup locations, several inconsistencies in TIST research accounts were noted, which made their way to a certain Washington journalist's desk. Similarities between TIST and Allies of Creation donors surfaced. Once the dots were connected, the story was compelling. Published in installments over a period of one week, the story sold.

With Houston as the latest major metropolitan area to succumb to climate change, the public wanted blood. Victor was a start, but there was a much better scapegoat—the NSF and the entire system of government-supported research. The spin-doctors of the far right made an emotional argument for doing away with the entire inefficient government research enterprise that not only had underpredicted the effects of global climate change but also was clearly corrupt from top to bottom.

***

Victor Hobbs's fall was cushioned by the short memory of the public and the influence of his associates. Following a lengthy investigation during which time a regular series of climate-based calamities dominated the news, he avoided conviction on bribery and fraud. He was able to plead plausible ignorance of the ultimate source of his funds. An unlucky postdoc from his lab at TIST took the fall for incompetent lab procedures that led to the spurious results for which Victor's benefactors had so dearly paid. There were many in Washington who wanted this matter resolved before anyone noticed that these same benefactors had been major donors to their campaigns. Victor relocated to Asia, where he easily found lucrative consulting work.

Taking down the NSF took time as well. When Eva was called in to direct the NSF after the dismissal of Victor, she immediately realized that the task ahead was hopeless. The best she could hope for was a controlled fall that would maintain a basic structure of the organization. Her hope now was to preserve a vestigial structure that might channel a future renaissance in science.

As Jill packed up her office, she carefully placed the 20-year-old photo into a box containing family photos, memorable knick-knacks, and her favorite coffee mug. Yes, there had been good guys and bad guys, but initially they had all been on the same side: the side of science.

Chapter 6

Message to Earth

Robert L. Olson
Institute for Alternative Futures

At 5:30 a.m. Marjorie Murray was beginning her morning routine of reviewing her goals—a practice she'd developed to keep from losing touch with them in a job with an inbox always stuffed with distracting problems. Because she was awake at that early hour, she saw the message the moment it arrived. The POTUS Communications System console squawked an alert and displayed the message. It was obviously not a message that should have gotten through to the president of the United States. She'd been hacked.

George Landau was paranoid, possibly clinically, but definitely by profession, since he was director of Central Intelligence. The Encryption Race triggered 40 years ago by Edward Snowden had half-blinded the intelligence community. The calls, texts, emails, and other data streams that were once routinely intercepted are now mostly indecipherable. Plus there are criminal organizations, radical libertarians, terrorists, and individual hackers trying to blind the government still further—US intelligence and military networks

Flash Forward: A Series of Futuristic Vignettes
Edited by Nora Savage and Anita Street

ISBN 978-981-4669-44-3 (Hardcover), 978-981-4669-45-0 (eBook)
www.panstanford.com

have long been under constant attack. And now someone has penetrated the POTUS Communications System, the most secure system of all.

Within hours of the hack, Director Landau met with the CEO of Visicomm, the firm that designed and built the maximum-security POTUS system. Landau was impressed that the man made no excuses and didn't hesitate to share his view that someone inside his company who knows the technology intimately probably did this to highlight a security flaw.

"I appreciate your frankness, and I wish you were right," Landau said. "But we've learned that at least four other heads of state were hacked in exactly the same way at exactly the same time, and they use different communications systems. This is not the kind of thing governments want made public, so I wouldn't be surprised if several other national leaders have also had their communication systems compromised. Which means this is far more serious than just an anonymous tipster in your company."

They agreed to immediately set up three independent teams—one in Visicomm, one at the CIA, and one in the Naval Network Warfare Command—challenging each of them to find security flaws in the POTUS system.

Two weeks later Landau reported to President Murray that none of the teams had yet succeeded. "We've had all the Visicomm employees who ever worked on the POTUS system take full-scale brain imaging lie detection tests. They've all passed. But Visicomm isn't really the issue any more. Our assessment is that every single nation's top leadership communications has probably been penetrated."

They discussed a series of cumbersome new procedures for trying to ensure the validity of all communications to and from the president. Then President Murray asked Landau for his best assessment of the whole situation. "What's the significance of the message? Who might be capable of doing this?"

"The message itself is silly and harmless, just a vehicle for displaying someone's capability. It's that capability that's obviously so dangerous. There are different theories about who might have the ability to do this, but we really don't know yet. Admiral Rowley at the Naval Network Warfare Command says it's impossible to trace where the message came from. One thing's certain, though: the perpetrators of this hack want to get our attention. They've done an incredibly brazen act telling leaders around the world that they're out there and they can do this. I'm betting it's the beginning of something bigger, and before long they will act again."

Director Landau was more correct than he possibly could have imagined. Two weeks later the message appeared simultaneously on every computer, phone, tablet, smart paper sheet, eye display, holographic projection unit, and virtual reality (VR) viewer in every country and in every written language on Earth.

President Murray found it hard to keep to her morning routine of reviewing her goals, hard to think about anything besides this incredible hack that revealed the vulnerability of virtually everything connected to the Internet. Who could possibly have the technological capability to do such a thing? What else can they do with that capability? Why are they doing this? What do they want? She had no answers and no idea how to respond to the situation.

She was very glad Jim Robinson was joining her morning briefing by Director Landau.

She regularly encountered bright people at the top level of government and the private sector, but her science advisor was in a completely different league. Until she met Robinson years ago when she chaired the Senate Science Committee, she had never imagined it was possible for one person to know so much about so many things. And he was in a position to mobilize many of the nation's best scientific minds to respond to this threat.

"Is there anything new to report?" President Murray asked Landau.

"Just that the world is going nuts. The media are playing this up as the 'Story of the Century.' The brightest ex-hackers we have working with us are awestruck at this feat and have no idea how it was done. Fanatics in every religion are claiming the message is from God. There's even an old song about the words in the message that's being played everywhere. It's crazy."

"Jim, has the team of scientists you convened had any insights into what's going on?"

"I'm afraid not. We're still completely baffled. We've pulled in technologists from across the private sector and they're all baffled, too."

President Murray leaned forward, hesitated, leaned back, but then leaned forward again and spoke. "There's something I've been thinking about that I'd like your reaction to. George, in your catalog of nutty things going on, you didn't mention that some people are saying the message is from extraterrestrials. It does begin with the words 'Message To Earth,' which makes it sound like it's coming from . . . outside. Is that a possibility?"

Director Landau stood and paced in front of the president, speaking emphatically. "Let's not get fantastic. What we're dealing with here is actually just a single technological capability. It's the same trick that got the message into the POTUS system, applied net-wide. It's certainly impressive. Someone is a step ahead of us, technologically, but not necessarily that far ahead."

"The question is," Landau continued, "who could do this? Admiral Rowley and I both think that only a nation-state has the resources needed to pull off this stunt. And the leading candidate is clearly China. They're ahead of us in artificial intelligence, and we think a highly advanced AI system would be essential for doing this sort of thing. And the Chinese have been infiltrating other nations' communications systems for over 70 years."

"What's your view, Jim? Is it nutty to think the message might actually come from extraterrestrials?"

"Extraterrestrials are a possible explanation. But I have to agree with the director; there are other explanations that seem far more likely."

US intelligence capabilities focused on determining whether China was the source of what had become known as the "Global Hack." Several members of Congress publically accused China. Tensions flared between the two countries when a Chinese ship collided with an American cruiser in the South China Sea.

Then, a month after the Global Hack, long, narrow clouds like jet contrails began forming all around the world. When the sky everywhere was filled with them, the clouds began breaking into shorter pieces, twisting and turning, and arranging themselves to form the message. So many copies formed that virtually everyone on Earth could see it, written in one or more local languages.

If the Global Hack was the news story of the century, this was the news story of the millennium.

Or at least it was until the next month, when phytoplankton bloomed across most of the surface water of the Pacific Ocean and arranged itself into the form of the message. Pictures from orbiting satellites vividly displayed the message written across the face of the planet.

President Murray and Robinson sat together looking out across the lawn and Pennsylvania Avenue to Lafayette Square, talking informally like the old friends they were. "Jim, I heard you gave a lecture at the National Academy a few days ago, where you said you doubted if the message posed a security threat and you thought extraterrestrials might actually be responsible for it. Have you really come to think that's what's going on?"

"I still have no real evidence about who or what is doing this. I gather even Director Landau is finding it hard to imagine that China or anyone else knows how to write with clouds and coax plankton to write on the ocean. That leaves God or extraterrestrials, or something else we haven't imagined. As a physicist, I'm reluctant to turn to supernatural explanations, although I can't rule them out completely. So yes, I think you may have been right when you entertained the idea that the message is from extraterrestrials. I don't have any better explanation."

"But why, Jim? Why go to so much trouble to impress such a simple thought on us?"

"I've been thinking a lot about that. Maybe it's exactly what we need right now."

"What do you mean?"

"Are you up for a bit of a rant on what I've been thinking?"

President Murray smiled broadly. "Jim, you know I'm always glad to hear your rants."

"OK, think about our biggest challenges. Pick any one of them."

"Energy and climate," the president said immediately, picking the challenge she had worked on the most throughout her career in politics.

"A perfect example. How do you think we've done in meeting that challenge?"

"Well, we certainly haven't been as successful as we could have and should have been. Technologies for using energy more efficiently and renewable sources of energy were delayed for decades because fossil fuels were kept artificially cheap. We subsidized them and refused to figure in their costs to our health, the environment, and the climate or the huge military costs of protecting our access to them. We've finally had breakthroughs to small-scale fusion and artificial photosynthesis, but these technologies have come too late to head off what's going to be a 7 or 8 degree Fahrenheit increase in the average global temperature."

"It's clear now that we could have had those technologies decades ago if we had invested more in R&D. The scientific and technical progress we needed was within reach—it was actually the easy part. What was hard was overcoming the short-term self-interest of the fossil fuel industry and working through all the needed changes in ideas, values, policies, institutions, business models, and all the rest."

But this is what I've been arguing for years. You knew perfectly well that this is what I'd say, Jim."

Jim Robinson smiled warmly. "Of course I did. I just wanted to have you establish the pattern that I think is there in all the largest challenges we're facing: technical advances are a critical part of the solution, but what's really hard is the human dimension of change."

"Think about the food situation. The skyrocketing food prices, malnutrition, mass migrations, and conflicts we're dealing with now didn't have to happen. A faster shift away from fossil fuels would have slowed the loss of glacier water. Greater use of water-efficient technologies for agriculture and more investment in desalinization could have prevented so many aquifers around the world from being sucked dry. Soils wouldn't be so depleted if companies involved in industrial agriculture hadn't slowed the spread of sustainable agricultural methods for decades. Harvests could be larger, despite global warming, if opposition to the use of genetically modified organisms wasn't limiting the use of drought-resistant crops. Technology for seawater agriculture to grow food without rain could be making a huge contribution if there weren't such huge subsidies for land-based farming. We can grow meat without growing animals and return huge areas of land to producing crops for people, but the meat industry is devoted to turning people against 'Frankenmeat'."

President Murray said, "It's the same for the ecological reconstruction efforts I've been promoting. The experimental Restorer Domes that are bringing back areas of rainforest in Brazil show what could be done using ecological design methods. And we could use analog species and synthetic biology to bring back many of the keystone species critical for restoring the structure of damaged ecological communities around the world. We could reestablish biodiversity almost as quickly as we destroyed it, if we were willing to pay for it. But we aren't.

Robinson began to speak, but President Murray interrupted, "You don't need to go any further with this, Jim. You've made your point. It's the damn politics of change that's so hard!"

Robinson locked eyes with his old friend, who he cared for so deeply. "The politics is hard: overcoming special interests, convincing the public to support needed actions, changing entrenched policies and institutions—it's all hard. And now this situation with the message has led me to believe that there's something else, something deeper, that needs to change, because it's the reason all these other changes are so difficult."

"What do you mean, something deeper?"

"This will sound way too idealistic for Washington politics. But I really think we need to grow from our adolescent short-term, me-first orientation to a more mature longer-term, we-first, planet-focused orientation. And that involves much more than a shift in ideas. We need a change in our hearts."

"And that's what you think whoever is behind the message is trying to bring about. A change so deep it helps motivate everything else. Do you think that's possible?"

"From what we've seen so far, I suspect whoever is doing this won't stop until they think they've succeeded."

A month later the Earth's atmosphere cleared of clouds, humidity, and pollution, so everywhere the night sky was ablaze with stars. Then, by some incomprehensible process of optics and atmospheric distortion, the stars seemed to move. The Milky Way's river of starlight broke into smaller, brighter streams that arranged themselves to form the words.

Message to Earth
All You Need Is Love.

Chapter 7

The Failed Life of Reverend Bayes

Pasky Pascual

Gently, quietly, he stroked the blackened toes of the Italian girl huddled in front of him. "Please, God, if there is a God," the English priest murmured to himself, "take this pain from her and bestow it on me instead."

Reverend Thomas Bayes gazed into the invalid's dark eyes. The affliction stared back at him. It had a name, which the priest kept to himself, knowing his neighbors in Tunbridge Wells would drive the young girl away if they found out. No other illness stoked mob hysteria like the bubonic plague, the Black Death.

As the sun sank beyond the oak trees bordering Bayes' property, it cast long shadows into the room. The priest sat beside the girl's cot as the world around him darkened. Within the gloom, the psalmist's words reverberated, "My God, why hast thou forsaken me?" Reverend Bayes had never felt so alone, so much a failure.

* * *

Flash Forward: A Series of Futuristic Vignettes
Edited by Nora Savage and Anita Street

ISBN 978-981-4669-44-3 (Hardcover), 978-981-4669-45-0 (eBook)
www.panstanford.com

The priest first came to Tunbridge Wells 26 years previously, in 1734. His parish was a playground for the wealthy. Every day in summer, public carriages from London would discharge tourists drawn to the hamlet's iron-rich, health-giving spring waters. Visitors hailed from all over Europe, from countries as far away as Portugal, Hungary, and Italy. Dressed in clerical garb, Bayes walked among the vacationing horde as the tourists hustled through the pubs and shops around the village spa.

"Good day," Bayes would greet the shopkeepers "Business going well, eh? May God's good fortune continue to smile upon you" Despite his inherent shyness, he thought it his duty to bring God to the marketplace, rather than having the gentle townspeople visit his priory. A young pastor, he staked his callow life on the existence of a benevolent God, a belief made tenable by his community's bucolic, harmonious conditions.

Tunbridge Wells had spawned numerous establishments catering to the recreational needs of Europe's privileged class. Women in meticulously embroidered gowns sat down with men, puffing on clay pipes, to while away long, leisurely hours playing whist and other games with cards and dice. The town dandies clustered together for whisky and billiards as they slurred empty conversation.

"Read the latest treatise from Blackstone, have you? Quite a lot to say on inheritances . . ."

"Saw Voltaire's play the other month. Rather liked it . . ."

Truth be told, Bayes did not care for these busy, summer months. He much preferred the quiet, slow months of autumn and winter. Cocooned within his priory, watching the oak leaves outside his window fade from red to brown to finally fall on frost-covered ground, he could dive into the first of his two obsessions to study the doctrine of chances or probabilities.

He learned about probabilities when he stumbled upon a book left in his church by a vacationer. Written by the French-born mathematician de Moivre, it sparked in Bayes the notion that many events—the length of a human life, the weight of newborn babies in London, the height of men in France—conformed to a bell-shaped curve.

To the holiday crowd, games with cards, dice, and billiards were mere diversions from humdrum, pampered lives. To Bayes, these games were fodder for him to think about the fundamental laws governing all creation. And understanding these laws would bring him closer, he believed, to his second obsession—to know God's love, as manifested in this world.

He spoke of these laws to a visiting cleric, Richard Price, as they walked past empty game halls on a chilly day in December.

"Cast a fair die and what chance have you to roll a six?" Bayes asked the younger man.

"Why, a one out of six chance of course."

"You are quite right. But what if the die were *not* fair? Or better yet, what if you did not know if the die were fair or not?"

"If I thought the die were fair, I would think the chance would still be one out of six."

"Yes, but supposing I threw the die one hundred times and it *never shewed* six. What would you think then?"

"I should think the die were not fair."

"*Yes*," said Bayes emphatically, grasping Price's elbow as he neared his main argument. "But if six shewed twenty times when I threw the die a hundred times? What are your chances then?"

"Twenty out of a hundred. But this is a puzzle for schoolchildren, Thomas."

"I suppose it is," said Bayes. He chuckled slightly. "If we were talking about dice."

"Well, what are we talking about?" asked Price in a peevish tone.

Bayes stopped walking and held up his hand to hush his companion. They heard someone walk toward them from around the street corner. In a low voice, Bayes continued speaking.

"If this person is a woman," said Bayes, "She is five feet tall. If it's a man, his height is five and a half feet."

The villager, bundled completely in a long cloak, turned the corner and almost bumped into them. Price could not tell the person's gender but noted that he or she was a head shorter than Bayes.

A genteel voice emerged from the wellcovered figure. "Oh, Reverend Bayes! How ye scared me!"

"Nothing to fear, Mrs. Dashwood, nothing to fear. Merely out for my afternoon constitution."

They watched Mrs. Dashwood in silence as she walked away.

"Richard, do you not see the order that dominates our world? God has woven Himself into the very fabric of creation. Using our observation and our God-given mental faculties, why should we not grasp His intent by calculating the chances of life's every event?"

* * *

How naive. How utterly naive he was back then, Bayes scolded himself. Beside him, the young Italian lass let out a soft whimper, which drew him back into the darkened room.

"*Si dorme, principessa,*" Bayes told her. "You sleep. Sleep will take your pain away, Benedetta."

The priest had known her family for 17 years, since before Benedetta was born. Her father, Signor Lupino, owned several fishing vessels in Messina, on Sicily's coast, just off the tip of Italy's boot He brought his family to Tunbridge Wells every summer because he loved the English countryside and he enjoyed his conversations with the unconventional cleric. Signor Lupino was an amateur mathematician who fancied himself the priest's equal in his understanding of probabilities.

"*Si, si,* is so obvious, *no*?" was the Italian's dismissive riposte whenever the English cleric engaged him in mathematical proofs. Signor Lupino had no use for theory; he was primarily interested in applied probabilities, specifically the doctrine of chances as applied to gambling.

"I bet the tails, *no*?" he explained to Bayes. He meant the tail ends of a bell curve, the unlikely events. It was a gambling strategy that worked for the Italian.

When wagering on the outcome of two rolled dice, most gamblers bet on the likeliest outcome, a seven, as there were six ways for two thrown dice to combine and yield that value: 1+6, 2+5, 3+4, 4+3, 5+2, and 6+1.

On the other extreme, there was only one way to obtain a two or twelve: 1+1 and 6+6. Because of the rarity of these latter events, gambling establishments would pay the lucky victor a vast sum of money to reward him for beating the odds. To Signor Lupino, these payoffs were worth wagering on rare outcomes.

"Always, I bet the tails," he told Bayes in confident tones.

There was a reason Signor Lupino first came toTunbridge Wells 17 years ago. Earlier that summer, a Greek ship had pulled into Messina's port. Signor Lupino had a cargo of linen on board that ship and had gone down to the pier to help unload it. As he supervised his men in loading his cargo on to his carriage, he noticed several Greek shipmen dragging dead bodies across the gangplank. He rushed to the water's edge for a closer look at the deceased.

There were lumps all over the dead men's necks, arms, and thighs. Some of these lumps had split open and oozed a combination of blood and pus. Lupino's eyes grew wide. He had seen these symptoms before. The Black Death had sailed into Messina.

"Carry them back to the ship, boys!" he shouted at the Greek crew.

To his own men, who were waiting for him at the carriage, he yelled, "Take the cargo out of there! Burn it. Burn it all!

Hopping on his horse, Lupino rode into town to notify the mayor of the plague-ridden ship. The mayor gave orders to torch the entire ship and to quarantine the Greek crew.

Lupino watched the ship burning in the harbor, the flames leaping up to stroke the night sky. The blaze cast an orange light around the harbor. A large, gray mass was moving quickly on the water's surface toward the shore. Upon reaching land, the gray blob dissipated into individual dots that scurried across the sand. Lupin's heart sank as he watched the horde of rats scamper into Messina.

A week later, the first death was reported. A day passed and three more deaths occurred.

Lupino understood the risks immediately. Messina was at the very tail beginnings of a pandemic that would continue to grow . . . and grow . . . and grow. He booked passage for himself and his family to go to England, where they would stay over the summer.

Summer waxed and summer waned. Year moved on to year. Benedetta was born and grew older. And the Black Death continued to ravage Messina. By spending summers at Tunbridge Wells and by fortifying the walls surrounding his large estate, Lupino managed to keep the plague at bay, away from his family.

By the time the pandemic had reached its peak, it had taken 50,000 lives in Messina. Then, like a horde of locusts that moved on after decimating an entire field of grain, the plague relinquished its hold of the city. Fewer and fewer deaths were being reported.

"W've reached the other side," Signor Lupino told Bayes with relief that summer. "W've reached the tail end.

Benedetta's symptoms began with a fever. When, at summer's end, she began vomiting blood, Lupino beseeched Bayes to look after his daughter.

"I cannot stay. Business affairs call me back to Messina. But Benedetta . . . she will never survive the trip back home. She must remain here. Please, my friend, my very dear friend, you must look after my Benedetta."

"Signor Lupino, you have my solemn promise," Bayes replied. "I will look after her as if she were my own daughter."

* * *

Bayes looked down at Benedetta and smiled wanly. Good, he thought, sleep has stolen her pain away.

He had once thought God to be benign and purposeful. Now, the only thing Bayes could perceive was the arbitrariness and capriciousness of a random world. A few months ago, Benedetta was a vivacious, beautiful 15-year-old with the promise of a full, happy life stretching before her. Now, a rare event had struck her. At the very tail end of the cataclysm from which her father had spared no effort to protect her, she had succumbed to a fickle twist of fate.

Bayes collapsed back into his chair. He cast his eyes around the room at all the notebooks on which he had scribbled his silly probability calculations. None of his research would ever be published, he thought.

He pulled out his handkerchief, pressing it against his mouth to stifle a cough. He let his hand drop, allowing his blood on the handkerchief to smear his chin. At age 60, Bayes could feel his life draining away.

The Black Death was taking him away from work the world would never know, away from a God he ceased to believe. "And yet," he thought, "I would accept all of this happily, if only Benedetta would live to join her family in Messina."

* * *

In 1763, Richard Price, named Baye' literary executor, found an essay among the papers written by his friend. Through a meticulous set of definitions, propositions, and calculations, the essay expounded upon a method to determine an even's probability, given observations on the event. Price sent it to an academic panel, writing an introduction to say that Reverend Bayes developed a metho

> *to shew what reason we have for believing that there are in the constitution of things fixt laws according to which things happen, and that, therefore, the frame of the world must be the effect of the wisdom and power of an intelligent caus; and thus to confirm the argument taken from final causes for the existence of the Deity.*

* * *

It was 20 minutes past midnight but Dr. Edmund Martinez remained in his cluttered office at the National Institutes for Health, in Bethesda, Maryland. A new batch of blood and urine samples had been flown in the week before. The lab had just emailed the data from these samples earlier that afternoon. Martinez discovered something startling in the numbers and he wanted to verify his calculations.

Dr. Martinez was an epidemiologist who was working on a team investigating the outbreak of the Ebola virus. Most people exposed to the virus died within weeks. Martinez was charged with discovering why a fortunate few managed to survive. Specifically, he was interested in finding out whether peopl's genetic makeup had anything to do with their ability to combat the virus.

He loaded the statistical package R into his laptop. Pulling up the data from the lab, he began running an algorithm called Markov Chain Monte Carlo, which enabled him to estimate probabilities according to the methods of Bayesian statistics.

Just 20 years ago, when Martinez was in graduate school, Bayesian statistics was an interesting, if theoretical, field of studies. But thanks to the advent of more powerful microprocessors and more efficient computer algorithms, Martinez could now implement Bayesian methods to study the genealogy of cohorts from all over the world.

Martinez poured himself another cup of coffee as he watched his laptop screen update the number of rows of data through which his algorithm was navigating. The data consisted of gene sequences in the mitochondria, the cellular component that produces chemical energy. Unlike the mix of paternal and maternal genetic material in the cell nucleus, mitochondrial DNA derives exclusively from one's mother. In theory, a perso's mitochondrial DNA contains the same sequence as his or her mothe's, as his or her mothe's mother, and so on. It can therefore be used to track the perso's maternal ancestry though hundreds of generations.

Through sheer happenstance, a mutation can occur if a sequence of mitochondrial DNA is not faithfully copied over to the next generation. By analyzing whether a given population shared the same mutations, Martinez could infer the degree to which people within a cohort shared common female ancestors. By assuming the rate of mutations was stable over time, Martinez could infer when a maternal ancestor might have lived.

Martinez sipped his coffee as the results of his analysis flashed across the screen. It was hard to believe but the data was clear. There was an unlikely degree of homogeneity in the mitochondrial DNA of his randomly sampled cohort from Messina, in Italy.

"It's as if," he told himself, "ninety-eight percent of all the people currently living in Bethesda could trace their lineage to a single woman who lived here in the 1750s."

He allowed his Bayesian algorithm to run through its final computations before turning off his laptop.

Chapter 8

Arachnophobia

Bruce Tonn

The steaming cup of Appalachian coffee did not disappoint—smooth to the taste yet bioengineered to provide a quick, forceful jolt. Outside, an orchestra of a dozen synthetic species of songbirds was performing an intricate and beautifully coordinated melody. The creatures inhabiting the house were going about their own business, weaving, transporting, digesting, and excreting. A slight vanilla smell permeated the air inside and out. Kate saw her reflection in the window and was again pleased to not look her age. But, with a big sigh, she acknowledged that this delicious, peaceful morning was at its end. Her husband of over a century was noisily muttering as he was making his way to the kitchen.

"Do I have to?" lamented Bob, as his flailing arms almost crushed a beetle crawling up the wall. A smile briefly crossed his face when he saw that a fresh and steaming cup of hot chocolate with community-grown cocoa was waiting for him, along with a plate of newly printed old-fashioned, double-stuffed Oreos.

Flash Forward: A Series of Futuristic Vignettes
Edited by Nora Savage and Anita Street

ISBN 978-981-4669-44-3 (Hardcover), 978-981-4669-45-0 (eBook)
www.panstanford.com

"Good morning, honey. And, yes, you do," replied Kate, his wife of over a century.

"I really have to participate in a Global News Network (GNN) documentary about me? About the space elevator and Probes Project? Our community? Our home? Why? Oh, and I am sorry I didn't say good morning. Good morning!" Kate thought that Bob actually turned a whiter shade of pale as he realized his faux pas. After so many years, why can't he learn to say "good morning" first, they both thought!

Kate's adopted Southern charm kicked in. "Bless your heart, Bob. Thanks to your advocacy for resettlements, and now the Probes Project, you have unwittingly become one of the most famous people in the world. And, because you are a dedicated, successful, and, frankly, paranoid introvert, there is almost nothing in the cloud about you. I know you do not keep track of the collective human consciousness, but I sense the Noosphere daily. Many people simply do not believe you exist. Some believe you are a ghost. Some believe you are a hoax, a fake persona created by the transnational backers of the Probes Project. Your anonymity will be protected, I promise: Your face will never be completely shown and your voice will be disguised through a translation system that will make you sound like a *Muppets* character or a character from that still scandalous animation series *Family Guy*. Sweetie, the world needs to hear your ideas directly and meet you, warts, spiders, and all."

* * *

A few hours south of Bob and Kate's home in Southern Appalachia, a documentary team from the GNN had gathered in Atlanta. "Bob really exists!" exclaimed Janet, the documentary team's young and enthusiastic intern. Among her favorite jobs was keeping track of the questions the documentary team planned to ask.

"Yes," replied Melissa, the documentary team's lead. "GNN upper management was mysteriously contacted and then somehow arranged to do an exclusive documentary on him. This could be huge for us."

Dan was worried. He was in charge of producing the real-time streaming component of the documentary. He was having a hard time imaging someone he could not picture, much less fathom.

"Nobody knows anything about him. He seems to be the only person on the planet with no cloud persona. Where does he live? Where was he born? How old is he? Is he really a he?"

Melissa raised her hands to calm Dan down. "We caught a break. We got a tip from Bob's Decision Day nemesis, the equally mythical Anti-Bob, that he actually lives in an innovative DNA community located in the midst of several resettlement communities in the Southern Appalachians. He has been in our backyard all this time." Melissa was clearly caught up in the intrigue of the situation.

Decision Day. Bob and Anti-Bob. The team was awestruck. That was the day the entire world voted as one to decide on whether to cooperate on a global resettlement plan to move tens of millions people from the areas hardest hit by catastrophic climate change. The vote was extraordinarily contentious. Regions to receive immigrants from around the world were particularly worried. Somehow, Bob's and Anti-Bob's voices rose above the cacophony. Their back-and-forth discussions—which were simultaneously bitter, well thought out, and considered—clarified the issues. Bob's passion, vision, and can-do attitude won the day in this worldwide vote. The decision directly led to numerous resettlements in places that had not been as ravaged by catastrophic climate change, like Southern Appalachia. Bob and Anti-Bob were also key in Decision Day 2, when the world voted a second time, this time to approve a global effort known as the Probes Project, to find earth life a suitable home beyond the solar system.

The resettlement process in the Appalachians was particularly tense. Those whose ancestors came to the region centuries ago were still fiercely independent and even after a century had not forgotten the government's taking of their land for dams for the Tennessee Valley Authority, the Great Smokey Mountains National Park, and the Oak Ridge Reservation's role in the Manhattan Project. Even before the resettlement process began, there were threats of violence. The bombing of the Bangladeshi resettlement, those displaced from their homes by sea level rise, was tragic but not a particular surprise. Melissa and her crew had covered that story nine years ago.

"Remember the resettlement specialist, Kate, who was on the scene, in charge, and successfully calmed everyone down? Her work paved the way for the peaceful resettlements throughout the United

States," Melissa continued. "Well, I contacted her to see if she knows anything about Bob and she does. She's his wife!" Melissa was clearly enjoying telling this story.

The fourth person on the team, Alice, was contemplating Melissa's story in the context of the documentary. Who are the subjects of the documentary going to be? What will the viewers want to see with respect to the daily lives of the documentary's subjects? Was this going to be a drama or a lighthearted farce or a trip down the proverbial rabbit hole? She was leaning toward a drama documentary, completely unaware that she was about to produce a modern version of *Alice in Wonderland*.

John, as usual, was represented by his holographic image. Melissa hated not knowing exactly where John was! He would be responsible for operating the fixed cameras and microphones and tasking the autonomous quad-propeller drones and the robotic tracking bees to follow promising story lines in real time. The holo John was rapidly churning through seemingly every color combination known to humankind. "She is married to Bob? She never said a word! Of course, who knew to ask her about him back then? And, Melissa, my little holographic image of you, floating I tell you not where, suggests there is one more piece to the puzzle. Let me guess—Kate's assistant, the one who was helping her that evening, the one who was pregnant, the one who gave birth right there in the Bangladeshi community center that night and, really, helped solidify the peace, she was her daughter, right?"

Melissa smiled. "You're almost right, John. Kate's assistant was her granddaughter. The baby is Kate and Bob's first great-granddaughter, Roberta." Melissa briefly wondered if Kate had somehow orchestrated the birth, which totally refocused discussions from intolerance to the celebration of life, in anticipation of some violent act. No, that seems quite implausible, thought Melissa. So she continued her report. "According to this note from Kate, they have taken to calling her Bob2, although it doesn't say why. Janet, please note that we need to ask this question. Anyway, Kate goes on to say that her husband was born in St. Louis and had a conventional suburban upbringing."

"That is so cool. We learned about the suburbs in our history class," Janet chimed in.

"Wait, how old is Bob?" Dan asked.

"We do not know for sure," Melissa answered. "He was one of the first to receive the diagnosis of futures vertigo after a major FV seizure while in college. As you know, FV was a mid-21st-century affliction, brought on by the combination of catastrophic climate change and the exponential change and convergence of advanced technologies. Our generation has no problems thinking long term, but back then, many people suffered tremendous mental distress when they tried. Anyway, we can assume he was probably around 20 years old in 2050, so he's approximately 130–150 years old now. His FV seizure turned his life around. He got a degree in sustainable development, and then advanced degrees in synthetic biology, engineering ecology, and human-environment systems design. He was one of the first re-environmentalization specialists to successfully integrate wildlife into human settlements in order to re-expand ecosystem ranges and to help save endangered species. Kate says he is known amongst his peers in the Appalachians as Spider-Man for his success in saving many species of spiders indigenous to the region." John thought he saw Melissa's holographic image flinch as she read that last bit.

"Bob is head of the life discovery team of the Probes Project. His position is an acknowledgment of his extensive knowledge of life gained from his innovative re-environmentalization efforts, a reward for his role in tipping the Decision Day 2 vote in favor of the Probes Project, and a result of the fact that the Probes Project is based at the old United States Department of Energy site in Oak Ridge, Tennessee. The first US space elevator will be located there, as will the United States' manufacturing facility for the 10,000 space probes that will be launched next year to explore possible new homes from humanity. We will tour this site as part of the documentary," said Melissa.

* * *

Kate continued prepping Bob for the day's upcoming visit. From many years of experience, telling Bob exactly what he needed to do was the only way to relieve some of his anxiety and reduce the probability that anything would go seriously wrong. The pained expression on Bob's face privately caused Kate no end of guilt. She hated her secretive life and manipulations, despite their obvious successes and benefits to Bob and humanity. It was unfair to focus attention on Bob and away from herself, but it had to be done. Maybe one day she would confess everything to Bob, to being Anti-Bob and more, but not today.

"Three things, dear. First, take the crew on a tour of the space elevator site and the Probes Project manufacturing facility. Second, take the crew on a tour of our community and then our home. Lastly, please try to be sensitive to the fact that none of them has been to a community like ours nor experienced a home like ours. As you know firsthand, people's reactions to our community and house can be unpredictable, intense, and maybe even painful. Remember when your brother, George, visited for the first time? He described the experience as falling through the proverbial Kurzweillian singularity. So you must keep an eye on them, understood?" Bob nodded in agreement.

"The lead reporter is Melissa. She was also the lead GNN reporter for the crew that covered the African resettlement bombing. She was there and reporting live when Bob2 was born. She is always well prepared with questions but is high-strung. So, please just try to stay calm. Focus your legendary mind on answering her questions as best as possible. She will probably have four other crew members with her. Again, let me emphasize, try not to let them out of your sight. And, Bob, please have our beloved family AI, TES, inform me when ya'll get to our house. I would like to greet the crew at the front door." Bob and Kate assumed, rightly, that TES was listening to the conversation. Bob assumed, wrongly, that TES would inform Kate when the guests arrived, even though he had not provided explicit instructions.

"OK, Kate, sounds straightforward enough, though I feel quite nervous and uncomfortable about this. You know I am not good around strangers. And there is so much I need to do to equip the probes with the ability to discover life on other planets. I have some more testing areas to set up in some caves and abandoned quarries."

"Yes, Bob, I know you are busy, but really, taking out a couple of days of your life for this documentary will not set this multi-millennial project back too much," replied Kate.

* * *

Melissa was just about finished. "Kate concludes her note with a few additional details. Bob is on the tall side, thin. He is an introvert but can be quite engaging when you get to know him. Despite rumors floating in the Noosphere from the Decision Day episodes, he is not a Peter Wiggins megalomaniac, nor is he an Ender Wiggins hero type. He is just Bob, who, I must warn you, has the distinctive focus of someone who has undergone neurocognitive restructuring."

The team, including John's floating hologram, just stared at Melissa, not only because they all thought that neurocognitive restructuring was a myth and had never met anyone who had had their brain's memory system cleaned up, but also because the procedure was only done on people over 120 years old. Melissa waved her hand to break the spell that had fallen over the team. "We will be fine as long as we stick together."

* * *

TES woke Bob up early on the big day. He and his AI had grown quite close during the lead-up to the first Decision Day. Bob thought of The Everything System as a she, and the AI did not mind using a woman's voice with him. She helped him craft his responses to Anti-Bob's challenges to the resettlement plan. Bob still did not understand why he was singled out or how this Anti-Bob person knew enough about him to call him out before everyone in the world. Somehow, though, his essential anonymity was maintained even as he became a global celebrity. Things did work out for the best. At his urging, and from the contributions of millions of others around the world, humanity voted on that day to embark upon the resettlement plan and then a few years later to authorize the Probes Project.

Both Bob and TES knew that this day was going to be difficult. He was being forced to leave his cocoon, to deal with strangers face-to-face. Kate said that if the documentary was well received, this would be the last time he would have to deal with so many strangers at once. This is because Kate and others could do all of the follow-up interviews once the world knew Bob really existed and that he really was quite busy! Kate gave him a kiss and a hug, as did his great-granddaughter, Roberta, who was spending a few days at their house with several cousins while the various sets of parents were visiting other DNA communities around the country.

The tour of the probe manufacturing facility went well. Bob answered the crew's many technical questions. Dan and John were particularly interested in Bob's strategies for outfitting and programming the probes to detect life. For example, Bob was using not only the DNA community but also his own home as a test bed for the probes' life-detecting systems. Scattered throughout the surface waters and streams were nonliving chemical reactions and very rare biological organisms, like the ancient bacterial mats. Many things in the community moved, from inert wind chimes and propellers to robotic bees. Even caves were outfitted with a mixture of organic and nonorganic systems. Soon a probe would be launched in a geostationary orbit above the community and tasked with determining what was alive and what was not. Everything was designed to train and test the probe's life-detecting algorithms. Bob was even happy to share his thoughts about the newly evolving field of terra forming, even though that knowledge would only be applied outside of the solar system tens of thousands of years in the future.

As he saw how impressed the crew was with the space elevator site and the probe manufacturing facility, Bob got over his nervousness. They soaked up both the technical information and the history of the place. Only Janet, who had a knack for history, remembered that the site was home to the first operational nuclear reactor and the facility that produced the radioactive materials for the atomic bombs dropped on Japan during World War II. However, the crew was acutely aware of the global resettlement program. Hundreds of millions of people relocated to more sustainable homes, including millions from around the world to the relatively sustainable southeastern United States. Bob was patient and persuasive as he explained that the resettlement program led to concerns that life on earth could not be sustained indefinitely, that new homes needed to be found elsewhere in the universe, and that the Probes Project was an essential step toward realizing this goal.

The crew was surprised to learn that Bob's community, the DNA community, coevolved with the numerous resettlements in the region. The DNA community contained like-minded people, in this case with a shared fervor for all things biological rather than a shared ethnic or national heritage. Like resettlement communities, the population of his community was about 200, following Dunbar's tenets about the optimal group size. Still, everyone was a bit apprehensive as they hopped in the autonomous GNN van for their tour of the community.

The tour of the community went well, too. The DNA community was certainly quite different: Everything seemed to be alive. The vanilla scent was quite pleasant. There was no concrete or asphalt. Bob showed them the central agricultural facility, which they affectionately called the Sprawl Farm. A complex system of fuel cells, photovoltaics, wind, and micro-hydro-generated electricity for the community and also provided food. An equally complex 3D printing facility was busy printing gastronomic delights at the molecular level at one end, medicines at the other, and household objects in the middle, with all of the inputs grown right there or being recycled from old products. An energetic carbon-Lego assembler was constructing a table. The team began to relax, which infused Melissa with a sense of foreboding.

* * *

Kate removed her ear buds in time to hear a scream and a thud from the playroom and the sound of the front door closing suddenly. "Oh, Bob! You promised to let me know when you arrived." After taking a few minutes to quiet down from the eventful tour of their home, Bob and Kate sat in the jungle room of their home. "Bob, tell me what happened."

Bob ran his hand through his thinning hair. "OK. Really, I don't know what happened exactly, but I'll tell you what I know. The tour of the space elevator site and probe manufacturing facility went well. They loved our community and the Sprawl Farm. I gave them a taste of that new chocolate pastry that was just printed—you know, the one with the mood-calming additive. Then we headed over here. They came in a big van but left their equipment inside because this was just an exploratory visit. They asked a lot of questions as we walked to our house. Why was this called the DNA community? Was the community truly self-sufficient? Did we live with wild animals? Was our house made of mud? Why did 200 people live in the enclave, and why was 200 a magic number? Did our children run around naked? Really, they had a lot of misconceptions about our community and house, so I was glad that this first visit was just to scope things out."

Kate interrupted. "Did they seem apprehensive? Had they had any de-sensitivity training? You know that even the most educated people have trouble dealing with their fears of animals and insects and spiders? Did their questions raise any red flags with you that this visit might not go very well?" Bob squirmed a bit and worked on relieving his head of some more hair. "Maybe I should not have fed them the pastry. Seemed to really calm them down. So, they seemed fine but in, ah, maybe in retrospect ..."

Kate could not contain her exasperation, "Oh, Bob, how many times have we been over this?"

* * *

About the same time, Melissa had finally woken up in the local clinic. Her team was around her hospital bed. She went from groggy to focused fury in a matter of seconds. "Where were you all? I told you to stay by me. I told you that we needed to stick together to get through this first trip. What happened to each of you? Did you even make it into the house?"

Janet was dirty and frazzled. "During the walk to the house, Bob was explaining the enclave's composting operations, specially designed creek beds for endangered mussels, fish-farming tanks, the revival of the American chestnut through gene drive technology, and biomass crops. I really wanted to see all this, so I lagged behind the others to walk around the enclave a bit more.

"I wandered around the back of the house and then out behind one of the operations buildings. I could see a huge pile of smoldering dirt. The smells began to overwhelm me. The rotting of the compost pile was apparent. Then, came a very strong fish smell. As I turned the corner, a big front-end loader dumped thousands of fish carcasses on top of another compost pile. I walked up to take a closer look. The woman on the tractor started waving her hands, getting more and more agitated. A moment later, hundreds of weird looking seagulls swooped down on the pile. They landed on me, knocked me down, and pooped on my hair. I hate birds and they hate me. I totally freaked out. Then a bear came out of the woods, followed by a bobcat and its two kittens. I thought they were going to eat me.

Then something that looked like a dinosaur came out of the woods, too, like it ran right out of Jurassic Park. I ran into the woods, through several large spider webs. I can still picture one of the spider's eyes, staring at me in shock! I also managed to roust a nest of very strange-looking bees, which scared the honey out of them and me. I still don't know how I made it back to the van."

* * *

"When we got to the house, there were only four of them. I really don't know what happened to the fifth person," Bob reported to Kate. "She happened upon the fish dump at exactly the wrong time," surmised Kate. "Oh," responded Bob is a very weak voice. "Continue, please," directed Kate.

"OK, well, we got to the house. I asked everyone to leave their shoes outside, like always. No one was wearing socks."

"I am glad you noticed that," said Kate, very sarcastically.

"Hey, no need for that! I had everything under control. The Jungle Room was neat, and everything was picked up. The music was already playing. They seemed to really enjoy the music. We had a few minutes to tour and then get out of the room. Everything seemed fine."

* * *

"OK, Janet. Sounds like you had a harrowing experience. But that is no excuse for wandering off. Anyway, if you had been listening more closely to Bob," Melissa lectured, "you would have known that those weren't seagulls; they were passenger pigeons brought back from extinction and given seagull genes to crave fish. And yes, that was a dinosaur, a miniature velociraptor, which was also brought back from extinction. And, I do believe that was a nest of robotic bees programmed to pollinate the DNA community's fields. OK, OK. Back on point. So, what happened next? We were in the front room, the so-called Jungle Room. Bob seemed intent on getting us through the room quickly. Then we headed toward the back of the house." Alice looked the most sheepish at this point, so Melissa targeted her for the next interrogation.

"The room was so interesting! Grass was growing on the floor. It felt so soft. The music was hypnotic. What was it again?" Dan interjected that Bob said it was his 12th chromosome. "Wow, I wonder what my DNA would sound like. Anyway, Bob seemed in a hurry to move us along but I wanted to stay just a second longer. The far wall appeared to hold many interesting artifacts. I wandered over to have a better look."

* * *

"I had plenty of time to get the visitors through the Jungle Room before the mist started. How was I to know that one of them would linger?"

"Bob, you just need to be more aware of what is going on? Did you ever look back or count heads?"

"No, I guess I was too involved in explaining our wonderful house to Melissa."

"Oh Bob," muttered a clearly exasperated Kate.

* * *

"What I thought were artifacts were dens and homes for creatures and the living creatures themselves. I reached out to touch one and it moved and then disappeared into the wall. Several others also disappeared. One ran down the wall and bored into the floor. Then it began to rain! Flying insects emerged from the walls. They started to get into my hair. I fell back against the wall. My hand got stuck on some sap. It was dirty and gross. I was struggling to pull my hand out, when a snake crawled across my bare feet. I began screaming hysterically, but because of the noise of the misting and because the doorways to the rest of the house were now closed, no one heard me. The adrenaline rush gave me the strength to pull my hand out of the wall but not the brains to tell me not to put my hand in my hair. For some reason, the front door just opened on its own, and I ran out of the house wet, dirty, and with my hand stuck to my head."

"So, Alice, that explains your new, ugly buzz cut and your dirty, hairy hand," remarked Melissa. "That also explains why Bob rushed most of us through that room while explaining the house's compartments for living creatures to me in excruciating detail."

* * *

"I was telling Melissa about the house. I didn't want them to be afraid of anything coming out of the Jungle Room during the mist. I think Dan understood; at least I thought he did. He seemed to understand that although the house was constructed of carbon nanomaterials, all the walls were custom-designed to either accept and house or repulse various species. He seemed to understand that this was accomplished by the use of pheromones in the soils, as well as colors, other smells, textures, humidity, temperatures, water and food availability, the sizes of passages, sounds, and even electrical currents. In fact, each room had its own special ecosystem. Then he asked where the bathroom was."

"What, you let him go to the bathroom without any special instructions or warning," exclaimed Kate!

"Like I said, he *seemed* to understand what the house was all about."

* * *

"Janet and Alice screamed and I fainted. Great! OK, men. What are your stories? Why did you abandon damsels in distress?" Melissa goaded.

Dan was next. "Melissa, I didn't intentionally abandon you. I just went to the bathroom. I really had to go!"

"Yeah, yeah, we all know about you and your bathroom habits. So, what could possibly go wrong in a bathroom?"

* * *

"I pointed him to the bathroom. I told him we had a composting toilet with a very new design. That he would enjoy the experience. That he should take his time. That after walking through the kitchen, we would head on to the back of the house to visit the playroom."

"In other words, you didn't give him a clue about what the bathroom is like?" questioned Kate. Bob could only shake his head.

* * *

"So, anyway, I really had to go. As I was running off, I heard Bob instruct me to try to tell the difference between the beetles and something else. Honestly, I was not paying too much attention at this point. Not to go into too much detail, but I hurried in, closed the door, and settled in. It was crazy but I swear it smelled of vanilla and wintergreen in there," mused Dan. John interjected that the pervasive vanilla smell was probably due to some bioengineering of digestive track microbes to produce sweet-smelling poo. The others mulled this over for a bit before agreeing.

"There were no windows or electric lights. Bioluminescent plants provided the light. Like the Jungle Room, this room seemed to have stuff all over the walls and ceiling. In fact, there were bugs and things everywhere. I am not usually afraid of insects. But, in a small plant along the far wall, there was some interesting activity. Good thing I had on my binocular glasses. I focused them on the scene across from me and began to watch. That was a bad idea. My worst nightmare, being eaten by a female, was playing out in front of my eyes. A big female praying mantis was eating her mate. Chomp, chomp, chomp. I kinda lost track of things at this point." It looked like Melissa was about to eat Dan!

Dan continued, "My skin began to get prickly and I began to feel nauseous. Then I felt something crawl across my foot. Then something else. I let my gaze drop without readjusting the glasses and looked down. Gigantic dung beetles and ants were streaming out of the bottom of the toilet, carrying my, well, you know what up the walls! Some of them even looked robotic! I freaked out but didn't scream, I want you to know! I jumped up, but forgot that my pants were around my ankles. I tripped and fell into the shower area. The shower immediately turned on but the first mist was not water but mites. I know this because my face was smashed against the floor and even though they are small, up-close I knew! They were there to eat my dead skin. Then a heavy shower soaked me and then, well, I passed out. The next thing I knew I was out of the house, with a sensation of being carried out of the house by many very little things, like Gulliver. Oh, my God, I probably was," Dan cringed as he realized what had happened.

* * *

"So, you never saw Dan again?"

"That's right. I went back to find him but he wasn't there. But he took a shower! Why would he do that?"

"Oh Bob, I can assure you that he didn't voluntarily take a shower. Anyway, you said you took them through the kitchen? I was in the kitchen and didn't see you."

"We saw you. You had your ear buds on. You don't like my 12th chromosome opus, remember? You were frying beetles and ants for our pet aardvark. There was a plate of goodies on the table. I didn't want to bother you, and yes, I forgot that you wanted TES to let you know we were in the house."

"Oh, my, Bob, I think I know what happened to John."

* * *

John couldn't stop laughing. The image of Dan being carried out of the house by a stream of insects with his pants around his ankles was just too funny. Melissa rudely interrupted. "John, what is so funny? Why did you run out the house? Why do you look green around the gills?"

"OK, OK. I admit it. I am no better than anyone else. But I could have died! After Dan scurried off to the bathroom of horrors, we continued on to the kitchen. Bob's wife, I take it, was in there, at the stove. She had some headphones on and didn't hear us come in. Who would in any case, with floors made of grass? Anyway, Bob didn't want to bother her so he just kept going. Like Alice, I lingered a bit too long in this room. It was fascinating. There were herbs growing in niches in the walls. There were grapevines growing up the walls. There seemed to be tree branches coming through the walls that held apples and oranges and lemons. How did they engineer that? Anyway, it was interesting and I was finally beginning to feel at home in this crazy place. On the table was a plate of fried chicken nuggets that smelled delicious. I picked one up, gave it a good look. I have always thought I am not as stupid as I look but, well ... Anyway, I ate it. Crunchy goodness. Another intoxicating taste, like the pastry. Then I ate another. The plate was stacked, so I had a third and a fourth. I was eying a fifth tasty treat when I noticed some beetles crawling up the wall of the kitchen in front of the stove. The wife reached out, picked one right off the wall, and threw it into the frying pan. A side door miraculously opened. I managed to make it to the edge of the woods before heaving."

* * *

"So, now it is just you and Melissa and you still didn't think anything was wrong?" asked Kate.

"No, like I said, I was deeply engaged in conversation with Melissa. She seemed really interested in everything. She was itching a mosquito bite and I told her that was good because she was probably just injected with a vaccine for the West Nile virus. She had a bit of a headache coming on, so I gave her a medicinal cookie to eat. She said she really wanted to see the playroom because she has two young kids whom she really adores. She is always interested in new play ideas for her kids," said Bob.

"Bob, you of all people, you who has spent a career in re-environmentalization, who is unwittingly internationally known for designing human–nonhuman interfaces, should have remembered that a large fraction of the population has serious phobias. You missed them all this time. I bet that first poor woman suffered from ornithophobia, a fear of birds. And the second one, who most certainly ran out of our, screaming, probably suffers from ophidiophobia, a fear of snakes, and probably from mysophobia, a fear of germs. The guy in the bathroom probably suffers from myrmecophobia, a fear of ants, which was triggered by the beetles. Didn't it occur to you that Melissa might suffer from arachnophobia?"

"She never mentioned spiders. Never asked why they call me Spider-Man. I purposely did not meet them with any spiders on my shoulders. I didn't show them the spiders under the awnings at the side of the house. So, please understand that I really had no idea," protested Bob.

* * *

"OK, Boss," said Dan. "What happened to you?"

"I couldn't divert Bob's attention from his monologue. He is clearly a genius. His focus is astounding, as expected. And what he was talking about is genuinely worthy of a documentary. The house and the community are, as you all have said, fascinating. After really preparing ourselves by taking as many weeks of de-sensitivity training, as needed, we are going back, no protests!

"Anyway, by this time, I was apprehensive. Given Bob's nickname, I had been dreaming of spiders every night for the past week and waking up in a sweat each time. Then, the day before yesterday, I saw a small spider on my daughter's shoe. I freaked out and she screamed. I beat it to a pulp with my portable umbrella. It didn't seem like a good sign, given our pending visit to the DNA community. I really didn't want to see any spiders during our visit, and I really wanted at least one of you to be with me at all times! I had not seen any signs of any spiders, but then again, you all had disappeared. "

Melissa continued. "After the kitchen, we headed toward the playroom in the back of the house. Bob was talking about how it was hard to live with the black bears. They need a large range and substantial food supplies. If they cannot find their natural foods, they will scavenge for food in dumpsters and garbage cans. This causes lots of problems, for the bears and humans alike. His community has devised paths, which travel over roads, under roads, over rivers, and over railroad tracks, for the bears to follow when they move from one protected area to another. The paths have invisible electric fields to help keep the bears on the paths and moving along. In addition, people are discouraged from hiking the bear paths. They have planted more acorn-bearing oak trees as bear food sources, both in the mountains and in other protected areas. We have personal experience with the fish carcass dump, too, right, Janet? They have developed a black bear educational website, too."

"Out with it, you are stalling," Dan interjected.

"I'm getting there. Bob also talked about what they are doing to deal with coyotes and how they are getting people to actually let bats live in their belfries. The big Appalachian resettlement, which has been managed by his wife, if you didn't know, provided huge opportunities to design new enclaves from the ground up to support re-environmentalization efforts. The newcomers have been receptive to green roofs and homes on stilts with fungus farms underneath. They understood the need to quarantine potentially invasive species. They came here with many fewer phobias than the current residents have," said Melissa.

"Then Bob mentioned his great-granddaughter, Roberta. I remembered the story of her birth, in the Bangladeshi enclave the night of the firebombing. There was a picture of her on the wall, heading to the playroom. She looked so sweet! Bob said that Roberta took after him more than Kate, his wife. Bob hoped that Roberta would eventually head up the Probes Project after his time on this earth, which is why some were calling her Bob2. Just as I asked him why he thought this was a good idea, he opened the door to the playroom. Roberta came running up to me. Covered, covered ..." Melissa again looked faint.

* * *

"What, Roberta was covered in spiders? And her cousins were, too? Oh Bob," Kate cried. TES, who again had been uncharacteristically quiet during this whole discussion, finally blurted out through an image in a painting in the wall, "And I recorded it all! What fun! I hope this material makes it into the documentary."

* * *

Once Alice saw the footage of the team's initial visit, and similar mishaps from subsequent visits, she scripted the documentary to include TES's material for a humorous rather than dramatic result. Although it took every ounce of persuasive talent she had to convince Melissa to approve this tack, it turned out to be the right approach: The documentary went viral. Interest in biologically centric human settlements increased, along with the numerous areas of science underlying these communities' designs. More people began incorporating the various ideas from the DNA community and Bob's home into their lives. Interest and support for the Probes Project also rebounded nicely. Lastly, and perhaps most importantly, more communities and homes smelled of vanilla!

Acknowledgments

I want to thank Dori Stiefel and Jenna Tonn for their comments on various drafts of this story.

Chapter 9

Are Ye Gods?

Frano Agera
(pseudonym of Nora Savage)

As the train rattled against the rails, it jolted Diasporra out of sleep. It didn't matter, since her sleep lately was replete with nightmarish scenes and haunting bouts of insomnia. She tugged on the old fleece blanket she had wrapped around herself in a vain attempt to ward off the chilly feeling of dread and terror that seemed to seep into her soul. Looking at the blanket a twinge of melancholy pervaded her and a sad smile briefly illuminated her face. Almost 10 years ago her mom used to wrap herself in this same fleece blanket as she lay on the bed. The fleece blanket to which her mom clung, tighter and tighter every day, as her breathing became shallower and shallower and her life slipped away. Diasporra recalled how the mom she knew and her love were transformed before her eyes into a human husk whose spirit was ravaged by cancer. Had her mom begun to view the fleece blanket as a shield, a magical protective armor against the hordes of malignant cells that had invaded her body?

Flash Forward: A Series of Futuristic Vignettes
Edited by Nora Savage and Anita Street

ISBN 978-981-4669-44-3 (Hardcover), 978-981-4669-45-0 (eBook)
www.panstanford.com

Was her mom's mind so fatigued by relentlessly waging a losing battle that it forced untenable delusions into her consciousness? Was the same type of mental deception occurring now to Diasporra? Was she, indeed, so weary from waging a losing war against thoughts at once hostile and seductive that she believed an old keepsake of her deceased mother would somehow right her topsy-turvy world?

Half frightened, half frustrated with herself, Diasporra quickly stood up and discarded the fleece blanket into the empty seat, careful not to let it fall to the floor. As she turned toward the aisle and started to walk to the front of the train, Reggie and Clive walked rapidly to meet her. Having been friends with Reggie for most of her life, she could tell by looking at him that something had happened that excited him and made his eyes dance. Something was about to take place, and the electric energy vibes she sensed from him were very strong. Reginald Dillen and Diasporra had gone to school together—since "First Tots," as kindergarten was called in Liboon Plain, and had continued through high school. They became instant friends at the start, each sensing in the other a kindred spirit. Reggie was the first one Diasporra confided in when she resolved to stop drinking from the small bottles of water provided in their high school during meals. She recalled that night's conversation so vividly. That night five years ago when she introduced Reggie to Clive in the deep recesses of one of the school's storage room in the basement. She remembered everything as if it had just taken place a couple of hours ago. Perhaps some of her intellectual capacity was truly her own, she mused, as she recollected the events of that evening.

"Aw, come on Dia, it's our only free time and there's a sweet new twittering campaign that I want to get in on. It's like the coolest one ever!" Reggie said as Diasporra nudged him down the stairs. "If I don't get in on it tonight I won't become one of the 'First 100 Boosters' and will be labeled a 'Dufster' with all the other Late Larrys. Can't this wait?"

He stopped short of the storage room door, their special meeting place for private conversations. "Look, we can sneak down here tonight after lights out, and I'll stay here all night talking about whatever you want. I'll even risk getting caught and having demerits placed in my school social and behavioral record." The Liboon Plain

Boarding School was quite strict concerning adherence to rules and procedures.

"Just let me go and join this new active scene, and later it's you and me, I promise, all night."

Diasporra turned a somber face to Reggie, holding his attention. "This is important, Reggie. Not just to me, not just to us, but to the world."

"Give me a break, Dia, everything is earth shattering to you," he groaned. "Tell me honestly, will this important issue still be in the same state tomorrow?"

She continued staring at him without altering her expression and without blinking. She coolly answered, "Probably. But I might not be."

That was enough for Reggie to hear, and he followed her without another word or sound of protest. As they entered the storage room and headed for a nondescript door in the rear where there was a small inner room, Diasporra spoke in low, even tones, "Remember when we discovered this place? How we thought it was perfect because it lies on the very border of the school's property. We called our conversations here the Brink." She halted at the inner room's door and took his hand. "It's time to go over, Reggie."

Upon entering the room, after his eyes had adjusted to the darkness of the room, illuminated by the moon and one lone street light outside the window, Reggie realized there was someone else in the room. He instinctively drew back, attempting to protect Diasporra with his body. However, as he peered closely at the interloper quizzically, Diasporra spoke, moving from behind him.

"Reggie, do you remember Clive? From our SA outing three months ago?"

The particular Social Activities, or SA, day about which Diasporra spoke occurred in early November. The class was taken in the school van to Praised. The SA instructor had spent an entire lecture on the origin of the town's name—how the founders who held strict religious tenets, labeled "religious zealots," had chosen a name conscious both of their faith and of their hope for the future of the town and its citizens. Musing during the lecture Diasporra recalled her mom saying "God desires spiritual fruits, not religious nuts."

The first impression Diasporra had of the town's inhabitants, however, had been anything but hopeful.

On that clammy, cold day in November, when the class had visited 8 of the 12 churches in Praised, she sensed an overwhelming sense of hopelessness, of extreme fatigue, of despair. The school SA event to the churches in town was a natural activity, and the students were well aware of the role that these institutions played in Praised. Since the origin of the town, the churches of Praised had served as focal points for the various philanthropic and social events in town. At first glance the churches appeared well kept, clean, and orderly. However, Diasporra realized that what she saw was actually signs of a hasty cleaning. Traces of cleaning solutions, damp and slightly oily to the touch, and the faint odor of disinfectant lingered in the air.

In addition, the furnishings showed evidence of infrequent use. The hymnals, pews, meeting room tables, and chairs, and even the sanctuary seemed to Diasporra as newly arranged and set up. For what? She wondered.

Initially Diasporra thought it was evidence of a well-maintained and cared-for space. But that was before she met Clive, a resident of Praised, before he opened her eyes to the truth of the town and its fate. Before she realized her role in this unholy situation.

The SA outings were part of the core curriculum. The students were taken during these classes to an impoverished neighborhood, where they performed community service of one type or another. The outings were coordinated by instructors with assistance from members of local charity organizations and took place annually. On this particular SA event the assisting charity organization was one that was chaired by Diasporra's stepmother, Dominiqua, who was very proud of her leadership role in this organization.

Dominiqua and her husband firmly believed that charity work was an important activity, as did all of the parents in the town of Liboon Plain. Although Diasporra enjoyed most of the activities and shared the interests of her parents, town, and school in viewing charity work as good, she often overheard students complaining. Remarks included how ridiculous and tedious they felt both their parents and the school officials of Liboon Plain—LP, as they dubbed it—were. Diasporra did not agree with the prevailing sentiment on the campus, and on that day in November, she was one of the first to arrive, opting to take her own car rather than ride in the school van. She even volunteered to carry some of the donation boxes in her car.

Of course, Reggie was with her and helped bring the boxes inside. He teased her during the drive about her eagerness to get there so early.

"Is there a secret rendezvous you plan on having with one of the church elders?" he playfully asked. "Oh, I get it, early in, early out. I am all for that," he added.

She recalled, after the welcome by the church officials, the short speech her stepmother gave, and the presentation of the donations, how both parents and students rapidly scattered away. Reggie was among them, catching a ride with Howard, a close friend who also drove to the event. Unlike Diasporra, Howard arrived about half an hour late and did not hesitate to flee when the SA event was concluded.

Afterward, as she was packing up the leftover agendas and cleaning up the refreshment tray, one of the Praised teenagers walked over to where she was standing. She could still see the casual stroll of his walk and the slight smirk on Clive's face. As he walked toward her she began to feel a bit uneasy and she was not sure why. She could not understand why an approach by one of the residents of Praised should bother her so. This was her eighth SA event, and she had met dozens of residents, and she had never felt so uneasy. Yes, she felt compassion for the destitute residents of Praised and often stayed long after the SA event had concluded, talking with residents, both male and female, young and old. Why then did she now feel an overwhelming urge to turn and run?

As Clive approached the table where she worked, she did not look up again. She nervously hummed some television advertisement jingle that was stuck in her head for the past few days, trying to keep her hands from shaking. As Clive reached her side, she nonchalantly picked up a napkin from the pile of unused ones lying nearby and began to swipe the surface beneath her with vigorous circular motions. So intent was his gaze upon her that she could almost feel heat transmitted.

$$Q = m \times c \times \Delta T$$

"Q equals mass times the specific heat capacity times the change in temperature." She automatically recited the equation for heat transfer to herself. Finally she looked up at him.

"I'm sorry, is there a question for me?" She managed to get this question out without stammering. He was silent. She tried to match his gaze, staring at him as he did her. She could only do so for about 20 seconds, after which she dropped her eyes back to the table and continued placing unused refreshment items into the boxes.

"What do Praised and Liboon Plain mean to you? Why were these areas of the city so named?" he demanded.

She looked up again at him, more confident now. Her knowledge of the history of her city was deep. The woman who held the original title to the land on which the city was built was named Sronna Liboona. She had her middle name changed to Praised because when she was four years old, she sang a solo in church, and the song was reputed to be sung so beautifully that everyone said, "Liboona sure praised the Lord." The land that she purchased decades later and on which the city was built named the two areas, thus, in honor of her.

As Diasporra began to relay this information to Clive with self-assurance and even a little pride, she saw the smirk start to reappear on his face and she stopped her recitation midsentence.

"Just what is so damned funny? You asked a question and I am providing an answer to it."

She wondered if perhaps he was one of those men who felt intimidated by intelligent women. The type who interrupted a woman whenever she appeared to be smart, especially if she seemed smarter than he was. These men, she knew, were also frightened of anyone who differed from them and could often be observed belittling or teasing others.

He took out a pen and wrote something on a piece of paper and handed it to her. Now he did not seem so cocky; now the smirk disappeared from his face. After reading what was written on the paper, Diasporra looked up at him and said, "What does this mean?"

Clive turned and walked away. "Wait a minute," Diasporra called again. "What does this mean?"

Clive called back to her, "Look on the other side," and he exited the building. As Diasporra tried to contain her anger she flipped the paper over and saw his name and telephone number and at the very bottom, "Give me a call if you want the truth."

Diasporra had resisted the urge to call the number Clive gave her for about one week. Then one Friday afternoon as she left the school grounds for the weekend, she dialed the number. He answered after the first ring and in a manner unlike any other person she had known. There was none of the typical greeting.

"Meet me at the coffee shop on the corner of Fourth, the one you usually drop in on Sundays on your way back to school" was all she heard.

"What are you—" she began.

"I'll be waiting," he interrupted. "One hour. Please be prompt." Dial tone 0: he had disconnected.

"What? I am not going to meet you anywhere," Diasporra spoke half to herself and half to no one in particular.

Diasporra was never sure why she went to meet him. Nor was she ever truly sure why her initial annoyance dissipated so quickly after that call. She felt instinctively it was important to go; it was one meeting she should make every endeavor not to miss. He intrigued her, certainly, but there was something else.

That meeting so many months ago. She did not recall the coffee nor whether or not she ordered her typical pastry. She only remembered the words—three words she would not be able to forget. Three words that changed the course of her life.

As she stumbled into Reggie, the nudge caused Reggie, who had stopped immediately upon seeing Clive, to knock over some old journals piled on top of a table in the room.

The school used the storage room as a place to store old journals until the contents could be copied on electronic media. These journals were from the olden days when there were paper copies. The ink was fading and the paper was brittle and difficult to handle. Diasporra often wondered why the school bothered with them. Science had certainly progressed. What could be so important about old science from 3015 pages so long ago, synthetic bioarticles that the world had progressed so far beyond.

Diasporra, Reggie, and Clive talked for several hours. Clive eventually let Diasporra do the talking because it quickly became evident that Reggie received with hostility any word that Clive uttered. Once Diasporra went through the information piece by piece, she paused, letting Clive show the evidence. There was another two hours spent convincing Reggie—a total of almost eight hours.

The train rattled and shook amidst cheers of joy and sorrow from the crowds outside standing along the tracks.

"How technology and science have deceived us," Diasporra thought. "Or have they? Have I deceived myself?"

As she struggled to comprehend the enormity of the situation she now faced, she tried to grasp the unfairness of it all. She recalled again the moment when she knew the truth of her origin, the truth her parents never shared with her.

Her thoughts went back to that first encounter with Clive. After he told her his version of the town's origin, they talked in the coffee shop until it closed at ten that evening. Diasporra questioned Clive vigorously, demanding proof, clarification, details. That weekend Clive took Diasporra to libraries and basements in various buildings located all around the two towns. A range of emotions swept through her—disbelief, confusion, anger, fear, dismay, sadness, indignation. In the end anger triumphed, with a dash of indignation, and she resolved to take action. Her first step was to bring Clive to the campus and have a conversation with Reggie. She had to inform him, she had to convince him, she had to get his input and gain his involvement. So she brought Clive to the campus late Sunday night, and leaving him in the storage room she set off to locate Reggie. That first meeting of Clive and Reggie had not gone as smoothly as she had hoped it would; neither had it gone as tempestuous as she had feared it could.

Reggie defiantly folded his arms across his chest and glared at Clive. Still looking at Clive, he addressed Diasporra.

"What is this all about, Dia? Why do I have to be here?"

Before Diasporra could answer they heard footsteps outside of the room.

The opening of the door startled them. The science instructor walked in and demanded to know what they were doing. There had recently been some pranks on the campus that involved minor destruction of school property. April Fools' Day was a week away, and in an effort to curb or prevent additional incidents, the faculty were asked to patrol the campus during pre- and postclass times.

As a result of this increased vigilance they were overheard by Dr. Amanda Winstock, who now stood in front of the open door.

Eyeing Reggie and Diasporra and glancing suspiciously around the room, she asked, "What is going on in here?"

"I'm just searching through some of the old journals for information," Diasporra said, "and Reggie is helping me. What's the big problem?" Diasporra smiled at Dr. Winstock. She was a favorite student of hers, and placing the journal she had managed to grab just before Dr. Winstock entered, she added, "Why the Big Bro act Dr. W?"

Dr. Winstock smiled back at her and said, "Glad to hear it. Although I can't imagine there will be much useful information in these dusty old things."

Dr. Winstock added, "I would hope that you would know better than to engage in foolishness Dia. I have always—" she stopped. "Wait," her gaze shifted from Diasporra to Reggie and back again. "Why are you down here so late at night?" She moved close to Diasporra and peered at her neck. Diasporra turned red and protested, "Dr. W, REALLY! We have beds, you know."

Now it was Reggie's turn to be embarrassed, and he shifted from foot to foot, looking at the floor.

"I want to read through a few of these old journals I found here. They are interesting even though they are old."

Dr. Winstock smiled again. "Sorry, I did not mean to imply anything. Please don't stay down here much longer."

She turned and closed the door, and Clive, who had been behind it, wore a mischievous grin. Looking at him Reggie forgot about his embarrassment and began to feel angry again. Diasporra managed to calm him enough to explain the situation and to plead her case for his participation.

Diasporra remembered as she spoke and described the situation. How unreal it felt to her and still feels even now. Who could have imagined such a scenario, such a crafted world, such utter societal machination.

"You were vaccinated. Right?" she began.

"Of course, everyone was," Reggie replied, shifting his gaze from Diasporra to Clive and back again.

"The vaccination included something else—something called nanobio. This renders the receiving child a bonus beyond health. It imparts intelligence. And only the children of Liboon Plain received this type of vaccination. The children of Praised received a different type of vaccination. One that does not impart intelligence but on the contrary promotes a servile mentality and—"

"Are you crazy? What the hell are you talking about, Dia? What kind of science fiction, paranoid dribble is this guy feeding you?" Reggie's eyes widened in disdain and incredulity. This is ridiculous and I have wasted enough of my time. If you and this, this—"

Diasporra grabbed his arm and pulled him closer to her.

"We have proof. Let me finish the story. Please," she pleaded.

"The development of nanobiotechnology enabled people to live longer. Cancers and other diseases, which formerly yielded high mortality rates, were basically eradicated. So—"

"And now you think medical advances are bad, and you are throwing health care improvements into this nonsense? How can improving global health and wiping out disease be a bad thing?" he sneered.

Diasporra sighed. "It was not a bad thing in and of itself. However, with population increases these advances placed an unbearable and unsustainable burden on this planet. So there was a decision by a circle of world leaders that only selected portions of the population would be allowed to breed. As it happened the poorest were left to do the breeding. Then from the pool of children the wealthy would select one or at most two children. These children were given the nanobio vaccination. The others not selected remained with their parents, and this portion of the population became workers. Before you begin to protest and refuse to listen or threaten to leave, let me make one call," Dia asked.

Reggie stood, shaking his head as she dialed. Diasporra used the FacetTime option on her cell phone. A female voice answered. It sounded oddly familiar to Reggie but he could not place it.

"Hi, Tia. Hold on for a minute."

Diasporra handed the phone to Reggie. He almost dropped it and stared in disbelief. The girl looked exactly like Diasporra. Diasporra explained that Tia was her twin who had not been selected by Diasporra's parents and who was a friend of Clive's. Diasporra explained how the doctor was supposed to take all twins not selected to another country so that there would be no such meeting, no such revelation.

Then Diasporra and Clive showed Reggie other evidence. How Clive's parents suspected something in the vaccination and managed to steal some from Liboon Plain and gave it to him. So Clive was very intelligent, and he was the one who started looking into the matter when his parents told him the story of the switched vaccination several years ago.

It was hard to believe that all that had happened just four weeks ago . . .

The train jolted Diasporra out of her reverie of the past and back to the present time. Clive and Reggie were now good friends. The past three months and the ensuing activities—rallies, twiters, instamobs—had served as a bond to unite them. Now all three were working together to bring about a better global society, to demand global responsibility to the citizens and to the environment.

The videos they had made went viral just like the tweets and twitters. The culmination of it all was the rally that they had organized and to which they were headed.

This event was expected to be the largest synchronized rally ever and was taking place simultaneously around the world, whatever the time. This was the final showdown. What would happen?

Clive and Reggie approached her excitedly, talking about the trillions of tweets and twitters they had received confirming the rally.

Diasporra smiled. She wondered if her mom would have been proud of her. As the train slowed she looked from one to the other. They held up their palms and joined them together. They inhaled deeply and prepared for the moment, the rally, the revealing of the truth.

The only other sound was the rattling of pill bottles and the sloshing of water inside plastic containers. The rightful possessions of the global citizen—no longer hoarded secretly to be reserved for the few, the wealthy, the self-appointed heirs of the benefits of the world.

Chapter 10

The Path

Irelene P. Ricks[a] **and Phoenix A. Ricks**[b]
(aka R. Marion Troy and Pixie Carlisle)
[a]*Johns Hopkins University School of Education (SOE)*
[b]*Carlisle Troy Book Trust*

He had been missing for nearly 20 years. It wasn't often that people from the Town of Chadwick commented on the whereabouts of their residents, and he was no exception. No one made reference to his absence as he walked through the town, passing by familiar landmarks and faces. The lack of responsiveness had nothing to do with quaint New England stoicism or discretion, but plain old-fashioned indifference to someone regarded as a stranger. The only people who remained in "TC" (as the townies called it) past high school graduation were the people who had nowhere else to go. They had played too hard in school and had failed to gain admission to the large state university (usually so forgiving of its native sons and daughters), so they had nothing to look forward to but a life of hard, dull, low-paying work in a town where the largest employers were a large convenience chain and high-tech maximum security prison. TC folks married their so-called high school sweethearts, had three kids, one dog, and a mortgage they couldn't pay off until they were practically ready to retire from either work or life itself. Idle gossip

Flash Forward: A Series of Futuristic Vignettes
Edited by Nora Savage and Anita Street

ISBN 978-981-4669-44-3 (Hardcover), 978-981-4669-45-0 (eBook)
www.panstanford.com

about residents in TC was a trade industry like movies in Hollywood or politics in D.C., but it is said that only TC's undertaker knows where the real bodies and secrets are buried. TC has always been that kind of dead-end town.

He noticed that the house he entered was as he had left it. It was encouraging, but still a little unsettling, to see how little had changed while he had been away. Eighteen years was a long time to be gone. He had been a man-child at the time—not quite 13. Other than his mother, Mariah, who had tried repeatedly to find him, no one had seemed to really care that he had abandoned TC any more than they seemed concerned that he had returned.

The door was unlocked in the old ways when people never worried about the Thought Scavengers. Most people kept their doors bolted in the New Day to avoid fighting with them—at least in person. He was glad he didn't have to knock or ring the bell, but he was upset that better precautions had not been taken to ward off possible intrusion.

He made his way down the long corridor and it was just as he had remembered. Even down to the old wallpaper and artificial flower bouquet in the tall crystal vase by the stairwell.

His mother was seated in the kitchen, shelling green beans. She had a look of placid contentment on her face that he hated to disturb. He noticed with a sudden pang that she was older now, but still not old. Her hair had a kiss of gray at the temples and looked attractive on her unlined face. She had been a teenaged mother, no more than 15 when she had birthed him, so the fact that he was now 30 to her 45 made them seem almost like siblings.

"Mada?" That had always been his name for her. Not "Mother" but "Mada." It was a joke between mother and son that he had accidentally, in childish babble, foisted the Hindu name of "monster" on his beloved mother.

Mariah turned and gazed upon her son. Her eyes lit up from within like an oven in winter.

"Josey?"

He smiled. She was the only one to ever call him that. His stepfather, Joe Carpenter (no one who knew him ever called him Joseph), had always used his stepson's proper name, Josiah, saying it had great meaning and should give a young boy a sense of dignity that he would have to grow into some day.

He wondered if he had grown into it now, at 30, in his return home.

Mariah rose slowly and opened her arms.

Josiah hesitated. He closed his eyes, trying to remember the last time he had felt the wind from an embrace. It was different with the Thought Keepers. No matter how enlightened they were, or how kind, they didn't have the one thing that he had craved for the last 18 years.

A human touch.

Mariah enveloped Josiah into her arms and he melted into a fragrant scent of warm vanilla and honey. He knew that it was the tea that she was drinking and the soft perfume she had always worn. He remembered everything as if 18 years had dissolved into 18 minutes.

"Where are the Triples?" Josiah questioned, once his mother had released him and returned to her seat. The triplets were five years younger than Josiah and very hard to handle—at least as he remembered his then seven-year-old brother and two sisters.

Mariah smiled, and a dimple flashed winningly at the question. She gestured for Josiah to sit and he obeyed without question.

"August, Summer, and Autumn are doing well in school. They have completed The Test, you know. It's required now of everyone who turns 25. The scientists figured out that is the cutoff."

The Test was designed to cleanse the mind of the types of impurities the Thought Scavengers preyed on—hatred, jealousy, bitterness, sorrow. TC was like most of the towns and cities in the New Day in its desire to maintain the lowest-possible level of Scavengers who hovered at the edges of humanity like unwanted vultures. It was the twentieth year of the New Day, and in 2040 the cities had figured out how to marginalize the presence of the Scavengers, who were adept at snatching victims with only the least provocation. Parents had learned to teach their children from birth that negative emotions were to be shunned. As a concession, the Scavengers rarely took children younger than seven to allow parents the opportunity to instill proper virtues, but by the age of eight if children refused to repress negative emotions they were fair, succulent game. For the unfortunate who had not learned to sublimate, if not completely eliminate perverse emotions, the Capital City's biotech company, ExCELL, had devised a mood-altering pill, Ambrosia. Side effects of Ambrosia included forced cheerfulness and spontaneous laughter (most notable in young women) that were startling but nowhere near as unwelcome as the consequences associated with being ravaged by the Scavengers. Young people through the age of 24 were forced to take the pill each day, and by age 25, those with the tendency to still fight against the proclaimed benefits of the pill (sustained happiness and random good thoughts) were forced to endure The Test. Not completely harmless, The Test had the potential to leave participants with minds empty of not just emotion but intelligence as well. Those botched cases, the victims of what was called Ambrosia Sickness, were the stuff of urban legend, but Josiah knew the myths to be true. The Thought Keepers had told him and they never lied. During his stay with them, Josiah had seen an Ambrosia victim kept behind glass and watched like a specimen. The young woman would sit like molded clay, staring into space with blank eyes. The only sign of life was her persistent blinking.

"So the Triples passed without problem?" Josiah asked casually, shivering internally at the memory of the nameless young woman behind glass.

Mariah nodded, her eyes cloudy at the recent memory of her children's endurance.

"They had to have The Test, Josey. They were difficult to control, even on the Ambrosia. They missed you and blamed Joe and me for your disappearance. They also blamed the Capital City and just about everyone else they saw. They were like three tornadoes around town, making demands and violent threats! We were so afraid that the Scavengers would get them that I put Ambrosia in their food and drink and even ground it up as a powder and mixed it in with Summer's and Autumn's bath salts. I was afraid they would come down with Ambrosia Sickness, but I had to do something! You have no idea what your leaving did to us —." Mariah stopped, aware that her words were like swords in Josiah's heart.

The fact that his brother and sisters had almost been lost to the Scavengers or deadly Ambrosia Sickness because of his departure was too much to absorb. Josiah's expression was drawn, thoughtful.

"Would you like some tea, Josey? I can warm up some biscuits. I could swirl caramel on the top the way you like."

"So nothing happened to them?" Josiah asked pointedly, not interested in banal conversation or distractions like food.

Mariah shook her head.

"Once they turned eight we kept them home-schooled and dosed up on Ambrosia. We made sure they were safe."

Josiah nodded, satisfied that his mother was telling the truth.

"But from eight until now is a long time to keep them in the house, Mada. How did you do that?"

Mariah got up and turned on the flame under the tea kettle.

"Are you sure you don't want tea? I do."

Josiah felt his unease deepen. His mother's desire for tea was typically the result of bad news or something disturbing that she wanted to mask behind a steaming cup of something good. It had been her strategy since he was a little boy and he saw that she was no different now than she had been then.

"What do you want to tell me, Mada?" Josiah pressed quietly.

Mariah sat back down and studied the bowl of unshelled beans. They were a lovely deep green and perfect in their asymmetrical pile in the porcelain container.

"While you were away, things changed, Josey."

Josiah frowned.

Mariah took in a breath and explained.

"Companies like ExCELL became more important than you remember. I think it was just a storefront when you were a boy, but then the Keepers invaded and put the Scavengers among us to enforce what they felt was for our own good. Anyone over the age of eight was a target and parents were all terrified—me included. The Capital City Leaders invested heavily in research on cell biology, bioengineering, and genetics so we could take care of ourselves and not be completely destroyed because of random bad thoughts. The Leaders thought if they invested in ExCELL they could intentionally design infants predisposed to joy, reducing the need for drugs like Ambrosia."

Josiah's frown deepened.

"Did that work? I thought manufacturing babies went out with Hitler. You shouldn't genetically modify children like stalks of corn. It's not right, Mada!" Josiah exclaimed.

His next questions were posed with a bone-deep weariness, as if he were wearing the weight of the world on his shoulders.

"When are we going to learn to do what is right? Are they well cared for?"

The tea kettle went off. Its shrill whistle was unsettling, but Mariah seemed unperturbed. She rose slowly and walked over to turn off the flame.

Josiah watched as Mariah added a tea bag to a cup. She poured the hot water and watched as the water turned a deeper yellow. The fragrance of vanilla hovered in the air. Mariah returned the tea kettle to the stove and sat down.

"No. The children came out too different. Not at all filled with joy like ExCELL had promised. And not only that, they possessed an intelligence that was almost frightening. The Scavengers didn't even wait the customary eight years to take action. They snatched them out of cribs and cradles without the usual Parental Requisition used to cart away the "Over-Eights-Filled-With-Hate"—as the Scavengers call them. Once they took the children—babies, really—there was a massive revolt."

Mariah wrung her hands, remembering the terror.

"No one had ever fought the Scavengers before—at least not since The Struggle 18 years ago with your departure. Because of this war, entire towns and cities were decimated. People hid underground and the leader of the Thought Keepers had to come back and declare a truce. Chicago and Los Angeles were obliterated. New York was badly damaged. ExCELL was put out of business, and none of the children that were created from that awful experiment exist today. At least, not that anyone knows. There are rumors that the Scavengers weren't totally successful and some babies were hidden out of sight. TC has been on probation ever since. But maybe that will be lifted now that you're back?"

Josiah felt his heart almost stop at his mother's query. He had been told by the Thought Keepers' Council that things had been bad at home, but he had never guessed it was that bad. His sojourn had cost his people dearly.

"Where's Joe?"

For the first time in the last few minutes, Mariah smiled.

"Joe has built some beautiful buildings, Josey. You should see the church he erected two months ago in Chicago as a way to return the people back to their city. Some people have called it his biggest masterpiece."

Josiah made no comment to his Mother's obvious pride in her husband's work. As the silence deepened between mother and son to an uncomfortable degree, Josiah offered:

"I'm glad for Joe."

Josiah's relationship with Joe was complicated by the fact that his stepfather had been kind enough to step up as a father and husband when Mariah had been pregnant with Josiah, another man's child. Although Joe never made any obvious distinctions between Josiah and the triplets, Josiah knew that only the Triples belonged to both Mariah and Joe. Joe had always said that he considered that all four children belonged to him and were equally deserving of his love and protection, and Josiah was pleased that Joe felt, and acted, that way.

Mariah sipped her tea, gazing at Josiah with an unreadable expression.

"I want to see the Trips." Josiah sighed, missing his siblings with a sudden ache. He realized that there were few people his age in town who remembered or cared about him. But he could count on the Triples to feel differently. They had always loved their older brother without reservation.

"Joe will be home later tonight, but don't wait up for him. He comes in late most nights, meeting the needs of his clients. There have been many demands for new buildings here and around the country and Joe can hardly keep up with everything. TC is growing and it is due to the Thought Keepers who have herded them here over the last six months. I didn't understand why this was happening until your return. People kept asking why they were being shepherded to a place like this but I think those questions are about to get answered. By you."

Josiah watched Mariah sip her tea but said nothing in return. His reaction to his mother's brief exposition about the goings-on in TC was subtle, but it was clear that he felt strongly about what Mariah had shared. He also noted that she had not directly addressed his desire to see his brother and sisters.

"So when are you going to host your first Town Hall, son?"

Mariah asked the question softly, but it was clear from her sharp look that this was not an easy sort of question.

Josiah stood up. He walked over and planted a quick kiss on his mother's forehead.

"I think I'll go for a walk, Mada. I'll be back later." His deflection was as swift as hers had been about the Triples.

Mariah watched Josiah stride down the long corridor to the front door. She sympathized with his desire to reacquaint himself with the old neighborhood and felt it was important for him to reconnect with his town and any old friends that might still be around. His 18-year adventure was something that not many people had questioned openly. Josiah had always been a strange young boy and his quick mind and sharp tongue had felt alien to the soft-spoken, slow-moving TC residents. At least it had felt odd before the first Visitation of the Keepers and Scavengers.

Remembering how strongly Josiah usually reacted to change, she worried about his foray into TC. What would be his reception? Some of the newcomers would not recognize him, but some of the old-timers who had grown up in TC and had resented Josiah's differences might not be willing to relinquish those feelings.

The phone rang and a recent photo of Joe popped up. He was a handsome man in his early forties. He was unsmiling but seemed kind in a nonthreatening sort of way.

"Josey's home, Joe. What do we do?"

Joe's face replaced his photo. His expression of concern matched his tone.

"I don't know, Ree. But we must do something and not wait for something to happen to us. Not this time. Not again." "Ree" was Joe's pet name for Mariah.

Mariah sighed deeply. She knew that Joe was every bit as worried and unsure as she was about the purpose, and wisdom, of Josiah's return.

* * *

Josiah made his way to TC's community center more by instinct than design. It was a straightforward building that had been constructed by Joe more than 20 years ago, but it still looked new.

Josiah attributed that to the fresh paint on the slate siding and shiny glaze on the sign that read "Town of Chadwick Community Center." In the far-right corner was an inscription, "Joseph T.C.," in gold lettering. Joseph always signed his work that way. It stood for "Joseph The Carpenter." There was nothing particularly motivating about the signage or the actual architecture, but it had been one of Josiah's most favorite places because it had been built by Joe and he had once played there as a happy young boy.

The door swung open as Josiah made his way up the pathway, and several adolescents tumbled out, laughing and shoving each other playfully in the way of boys that age. They separated as they got closer to Josiah with three on one side and three on the other. One of the boys turned and looked at him. His eyes were neither friendly nor hostile, only mildly appraising. The door closed behind the group of boys. Josiah could see his reflection in the glass. His body was tall and lean. If he had been a woman, he might have been called graceful.

"You're new here," the boy stated flatly. It wasn't a question, because he had made it his business over the past 16 years to know everyone in TC.

"I was here before you were born and now I've returned," Josiah responded simply. He continued his forward momentum to the door and opened it.

One of the boys hooped at this declaration.

"Man, who in their right mind comes *back* to TC?" He asked the question to Josiah's back, not anticipating an answer, but receiving one anyway.

"I have unfinished business." Josiah tossed his response without turning, entering the building without another glance at the young boys. He had told them all they needed to know. And even that might have been too much.

The interior of the community center was simple. Oxygen masks hung from the walls as a reminder that the effects of the Ambrosia were sometimes too much for some children, particularly girls. He wondered if the researchers had bothered to test for sex differences in their studies before dosing the children indiscriminately. In the case of a bad reaction, it had been discovered that doses of oxygen diluted the effects of the drug, calming the hallucinogenic terrors of the youngest children.

"And so he returns!" This was called out by a voice Josiah recognized but couldn't see.

August! His brother, August, was here!

Josiah looked around the room but didn't see anyone. Not even the youngest child was apparent. It was as if his brother's voice was part of his own hallucination. Maybe the air was spiced with Ambrosia for the children as the town's way of providing extra dosing without the need for oral therapeutics.

"Gus!" Josiah called out gaily. "Where are you, brother?"

Josiah's nickname for August was "Gus." Autumn was "Tummy," based on his sister's indomitable appetite. He had called her that as a child and the name had stuck. But only family used it. No one else ever dared. Summer didn't have another name as she had always been something of an enigma to Josiah. She rarely spoke and even more rarely laughed. In many ways she was more Winter than Summer.

August ran out from one of the rooms, a sharp dagger in his hands, pointed at his older brother. Josiah instinctively threw up a body block and easily knocked his younger brother to the floor. Josiah retrieved the fallen knife and tucked it in his pants pocket.

"Is that any way to say 'hello' in TC these days?" Josiah didn't mean to ridicule August, but he was taken aback by the violent welcome.

"Thought Keeper!" August spat with as much venom as a betrayed spouse.

Josiah knew that 18 years was too long to be away for a brother of seven years of age. However, he had been unable to leave the Thought Keepers one second sooner than they had bartered. His time was not his own. He knew his brother was jealous that he had been selected, but that was also out of his control.

"Can we talk, Gus? As brothers rather than as adversaries?"

August sat up. He looked up at Josiah from the floor.

"What are you doing home, Josey?" August's voice softened and he used their mother's name for Josiah without thinking.

Josiah held out his hand to pull his younger brother from the floor. The two embraced quickly. Josiah pulled back, studying the man his brother had become.

"I came back to teach, Gus. To learn, to teach, and to change our world. I would like your help, but I will not wait for it."

August laughed at this and bowed low.

"When have you ever waited, Josiah? For me or anyone else?"

Josiah turned away from August with an expression of pain on his face and in his heart.

* * *

The small auditorium in the community center was packed and the podium on the tiny stage was dressed with a microphone and a glass of water.

Josiah listened to the quiet murmuring in the crowd and spoke when the rustlings were no more than sporadic whispers.

"My name is Josiah Carpenter and I am here to bring you news from the Thought Keepers."

The room fell even more silent at this proclamation.

"What do you—and they—want from us? Haven't we done enough? When is good, good enough?" This harsh inquiry came from one of the Elders.

Josiah cleared his throat. He had never disrespected an Elder and he was not about to start now. He had been known to challenge them, yes! But outright disdain or contempt for men who were much older was not an idea he had ever entertained.

Josiah recounted a story from his past.

"In this room many years ago, I defeated the Elders' arguments regarding theories you held as scientific truths. Remember bickering about the principles of chaos theory and the unpredictable nature of time and its elements that you touted as Gospel? Many of you disagreed with me then and challenged me on the propositions of *Lyapunov* time. You were wrong and I was only half-right. None of us is completely right about what we once believed, but I was only 12 and still untutored, so I had a legitimate excuse—what was yours? No one here knows the truth, but now I am that proof. I have visited the Thought Keepers and can share their knowledge with you. My time with them was a fraction of an hour, but by your time, I've been away for 18 years."

This last observation created more than a stir in the room. There were angry mutterings in the section that seated the Elders, and stifled cries of outrage. “How can that be?” was heard over and over mixed in with “Liar!”

Into this heated atmosphere Joe Carpenter entered. His masonry tools hung from his workman’s belt that he unbuckled as he walked. A hush fell over the crowd as Josiah’s stepfather made his way down to the front of the room to a seat marked “RESERVED for JC.” He sat down without saying a word and settled the belt on the floor next to his feet. A brief smile was exchanged between the two men and Josiah continued.

“The Thought Keepers want to tell you that we are doing better than before but there is much improvement to be made. As you sleep, your beta thoughts are relayed to them and are negative rather than peaceful. They want an answer to the question of when we are going to seek peace rather than war.” Josiah stated this dispassionately, but the reaction was not reciprocated. There were shouts of “Sleep Snatchers!” and “Dream Demons!” that echoed throughout the room.

Josiah continued as if there were no interruptions.

“I come with new technology and a strategy for harmony, my friends. I also come with a message that you must believe.”

The room fell silent once again.

“You LIE!”

This shriek came from a young man who appeared to be about 26 or 27. He was seated next to Summer and Autumn. He was holding Tummy’s hand, who yanked her hand away as he made his exclamation. She folded her arms and glared at the young man in defiance.

Josiah cleared his throat. His eyes were moist with emotion at his sister’s silent defense.

“If you don’t change, you will die. Do you understand this? The Thought Keepers told me that our planet is near an end, and they want to help us. But to help us we must become One. There can be no separation. Do you understand what this means?” Josiah pleaded.

The room was filled with a sudden electricity. It was no small matter to be told that death was imminent. Even if it was not entirely believable. And to be told that they must become One? What did that mean?

Josiah saw the fear in his mother’s eyes from across the room. He tasted her fear in his own mouth. He swallowed down the regret that this news had to be told by him.

"Who are you, Josiah Carpenter, to tell us that we are about to die?" This challenge came from one of the Elders.

Josiah met the eye of this Elder and held it.

"Who do you say that I am?" He asked quietly, so quietly that he could barely be heard.

At this, Joe stood and faced the troubled crowd.

"Josiah is my son. That is who he is. And he tells the truth. Please listen to him before it is too late."

Joseph was well respected, and the people, particularly the Elders, did as he said.

They stopped to listen to Josiah, to hear his words, and to understand his meaning. With an open heart, Josiah explained to his township what he had heard and seen over their past 18 years. But most importantly, he told them what they needed to know to live. He told them how there must be a complete eradication of all negative emotion, thought, and action. He told them about the process of neuronuclear genesis—ridding the human mind of all evil thoughts, taking us back to our primal state of goodness, taught to him by the Thought Keepers. He instructed them on the process of neurofusion, in which they would all fuse as One entity, One mind, and One heart. In this form, they would leave the planet to reside in a new and better uncontaminated place. A place without war, hatred, or fear. A planet doomed to self-destruct.

"But the Thought Keepers never told us that we were in danger. They just tried to make us better!" This rebuttal came from an Elder, but it was said with much less passion and more hesitancy than before.

"They've used violence against us! How does that support their plan of peace? People in our cities died fighting the Scavengers! Why would something good use evil to justify its end goal? That is not right!" Another Elder argued.

Josiah nodded.

"It's the reason for my return. It is my role to teach you to accept what you need to know. It is hard to believe this, but no one really died as a result of their so-called attacks. You will see what I say is true. There is not a lot of time, but there is enough if we start now. Are we ready to clear our minds and let the process of neuronuclear genesis take place? It will help us. I promise you."

There was another general buzz and arguments could be heard about Ambrosia, the Scavengers, trust, and Josiah.

"So, how will they change us, Josiah?" August stood and asked. His tone was simply curious, not confrontational as before.

"For those who are uncertain, they will come to you as you sleep and they have given me a device to use that will block your fears. For those of us who are unafraid, they will reveal themselves to us tonight without a shield. The device is not just for TC but for the world. I have been everywhere else, spreading the word. I came home to rest with you. You must trust me."

One of the Elders stood and rubbed his hands together nervously.

"Is there any other way, Josiah? I'm not sure that I like this idea. Becoming One. I did that with my wife and it didn't quite work. I think we're still Two."

There was muted laughter at this.

Josiah shook his head.

"I'm afraid there is only one way and you need to follow me. If you don't . . ." Josiah let the rest of the missing subtext speak for itself.

Joseph rose again.

"I'm ready, Josiah. Just say the word. I will follow you."

Josiah smiled. He knew that if all else failed, he could always count on Joe.

"I will let them know that tonight is the night." Josiah sounded more certain than he felt. He realized that there was still the question of free will and choice. Would the people of TC, and beyond, make the right choice? He knew that the final decision was out of his hands. The only direction he could lead them on was the path that he, himself, had been told to follow.

Index